Multimodal Signal Processing

Theory and Applications for Human–Computer Interaction

EURASIP and Academic Press Series in Signal and Image Processing

Series Advisor
Sergios Theodoridis

Titles to include:
Duhamel and Kieffer
Joint Source-Channel Decoding, ISBN 9780123744494

Thiran, Marqués and Bourlard
Multimodal Signal Processing, ISBN 9780123748256

Multimodal Signal Processing

Processing

Theory and Applications for
Human–Computer Interaction

Edited by

Jean-Philippe Thiran
Ferran Marqués
Hervé Bourlard

AMSTERDAM • BOSTON • HEIDELBERG • LONDON
NEW YORK • OXFORD • PARIS • SAN DIEGO
SAN FRANCISCO • SINGAPORE • SYDNEY • TOKYO

Academic Press is an imprint of Elsevier

Academic Press is an imprint of Elsevier
The Boulevard, Langford Lane, Kidlington, Oxford OX5 1GB, UK
525 B Street, Suite 1900, San Diego, CA 92101-4495, USA

First edition 2010

Notices
Knowledge and best practice in this field are constantly changing. As new research and experience
broaden our understanding, changes in research methods, professional practices, or medical
treatment may become necessary.

Practitioners and researchers must always rely on their own experience and knowledge in
evaluating and using any information, methods, compounds, or experiments described herein. In
using such information or methods they should be mindful of their own safety and the safety of
others, including parties for whom they have a professional responsibility.

To the fullest extent of the law, neither the Publisher nor the authors, contributors, or editors,
assume any liability for any injury and/or damage to persons or property as a matter of products
liability, negligence or otherwise, or from any use or operation of any methods, products,
instructions, or ideas contained in the material herein.

British Library Cataloguing in Publication Data
A catalogue record for this book is available from the British Library

Library of Congress Cataloging-in-Publication Data
A catalog record for this book is available from the Library of Congress

ISBN: 978-0-12-374825-6

For information on all Academic Press publications
visit our Web site at *www.elsevierdirect.com*

Typeset by: diacriTech, India

Printed and bound in the United States of America
10 11   11 10 9 8 7 6 5 4 3 2 1

Working together to grow
libraries in developing countries

www.elsevier.com | www.bookaid.org | www.sabre.org

ELSEVIER   BOOK AID
                International   Sabre Foundation

Contents

Part II
Multimodal Signal Processing and Modelling   143

Mihai Gurban and Jean-Philippe Thiran

Konstantinos Moustakas, Savvas Argyropoulos and Dimitrios Tzovaras

Andrei Popescu-Belis

Over the last 10–15 years it became obvious that all/most human–human interactions, especially in the context of collaborative and creative activities, necessarily involved several communication senses, also referred to as 'modes' (hence the term 'multimodal'). Complementary, often asynchronous (not carrying the same semantic information exactly at the same time), and interacting at different hierarchical processing (granularity) levels, these very sophisticated channels for communication and information exchanges include, amongst others, spoken and written language understanding, visual processing (face recognition, gestures and action recognition, etc.), non-verbal cues from video and audio, cognitive psychology (emotions), attention focus, postures, expressions. Human-to-human interaction is also constrained by implicit social rule parameters such as group behaviour and size, level of common understanding, and personality traits of the participants. Still, even if communication is not always optimal, humans are adept at integrating all those sensory channels, and fusing (even very noisy) information to meet the needs of the interaction.

Machines today, including state-of-the-art human–computer interaction systems, are less able to emulate this ability. Furthermore, it is also sometimes claimed that progress in single communication modes (like speech recognition and visual scene analysis) will necessarily rely on progress in those complementary communication modes. Based on this observation, it became quite clear that significant progress in human–computer interaction would necessarily require a truly multidisciplinary approach, driven by principled multimodal signal processing theories, aiming at analyzing, modelling, and understanding how to extract the sought information (to meet the needs of the moment) from multiple sensory channels.

The research areas covered by the present book are often at different levels of advancement. However, while significant progress has recently been made in the field of multimodal signal processing, we believe that the potential of the field 'as a whole' is still underestimated and also under-represented in the current literature.

This book is thus meant to give a sampled overview of what is possible today, while illustrating the rich synergy between various approaches of signal processing, machine learning, and social/human behaviour analysis/modelling. Given the breadth of this book, we also decided to reduce its technical depth, hence making it accessible to a larger audience, whilst also showing its potential usefulness in several application domains.

Several funding agencies recognised early that multimodal processing theory and application development had to be harnessed broadly, and several major projects were initiated. It is fair to acknowledge that several contributions to this book are the result of some of those projects, including the Swiss NSF 'Interactive Multimodal Information Management' (IM2, http://www.im2.ch) National Centre of Competence in Research (NCCR), the European FP6 Integrated Projects on 'Augmented Multi-party Interaction' (AMI), 'Augmented Multi-party Interaction with Distance Access' (AMIDA, http://www.amidaproject.org) and 'Computers in the Human Interaction Loop > (CHIL, http://chil.server.de), as well as the European FP6 Network of Excellence on Human–Computer Interactions (SIMILAR, http://www.similar.cc). But, besides funding, this book is also the work of many individuals whose common goal is to advance the knowledge in this complex, but key area, of multimodal signal processing, and we take this opportunity to thank them all for their high-quality contribution.

Hervé Bourlard, Jean-Philippe Thiran and Ferran Marqués
Martigny, Switzerland
10 September 2009

Chapter 1

Introduction

Jean-Philippe Thiran*, Ferran Marqués[†], and Hervé Bourlard**
*Ecole Polytechnique Fédérale de Lausanne, Switzerland; [†] Universitat Politècnica de Catalunya, Barcelona, Spain; **Idiap Research Institute and EPFL, Switzerland

A multimodal system can be defined as one that supports communication through different modalities or types of communication channels. In general, multimodal systems are also considered to use concurrent processing and to perform fusion of multiple, possibly asynchronous, input streams. For example, a framework for multimodal human–computer interfaces can be described as using a combination of modes (e.g., languages), channels (audio, video), media (speech, text, sound, graphics) and styles (menu, natural language, windows, icons).

Although multimodality has been discussed in research and science for several years, the computer science community is however just beginning to understand how to design well-integrated and robust multimodal systems. The proposed area of study is vast, covering disciplines such as computer science, engineering, linguistics, cognitive sciences, human–computer interfaces and psychology. This book however focuses on the signal processing and machine learning aspects of the area, hence mainly addressing specific (non-exhaustive) approaches of audio and video processing, joint processing (fusion and synchronisation), coordination and indexing of multimedia and multimodal signals or data, typical multimodal applications and related database architectures.

This book is thus a survey of the state of the art in a large area of topics, from video, speech and language processing to multimodal

Multimodal Signal Processing, ISBN: 9780123748256

processing, human–computer interaction (HCI) and human–human interaction modelling. The applications of signal processing and statistical machine learning techniques to problems arising in these fields are the two major themes of this book. Assuming basic knowledge in those areas, but given its broad nature, the goal of this book is to provide the interested reader (e.g., Master and PhD students, researchers in R&D centres and application developers) with an overview of the field, the capabilities and limitations of current technology and the technical challenges that must be overcome to implement multimodal interactive systems.

All contributors of this book are recognised for their expertise in the field and have been involved in several large scale projects targeting the development of complex multimodal systems.

This book is organised in three parts.

Part I, entitled 'Signal processing, modelling and related mathematical tools', gives an overview of the elementary bricks involved in multimodal signal processing for HCI. As such, this part is therefore mainly unimodal. The reader will find here an introduction to 'speech processing' (Chapter 3), including sections on speech recognition, speaker recognition and text-to-speech synthesis, as well as to 'natural language and dialogue processing' (Chapter 4). An introduction to image and video processing is given in Chapter 5 in the context of HCI, i.e., focusing on the main components used in multimodal HCI systems, such as face analysis or hand, head and body gesture analysis. Finally, handwriting recognition is introduced in Chapter 6 as another modality frequently involved in multimodal HCI systems. These chapters are preceded by an overview of the main machine learning techniques used in multimodal HCI (Chapter 2).

Part II is dedicated to the presentation of technical works recently developed in multimodal signal processing for HCI. First, the concept of multimodal signals and multimodal signal processing is introduced in Chapter 7. Then, the key problem of 'multimodal information fusion' is addressed in Chapter 8, detailing the most successful types of fusion schemes. Chapter 9 gives a first practical illustration of multimodal fusion, in the typical case of audio and video streams, with application to audio-visual speech recognition and audio-visual speaker detection. Chapter 10 provides a second perspective on this problem, addressing the concept of modality compensation and audio-visual fusion for disabled people and biometric identification. Finally, this part is concluded by Chapter 11, addressing the important aspect of managing multimodal data. This chapter outlines the main stages in multimodal data management, starting with the capture of multimodal raw data in instrumented spaces, discussing the challenges of data annotation, analysing the issues of data formatting, storage and distribution, including the access to multimodal data sets using interactive tools.

The last part of this book, Part III, presents several illustrations of multimodal systems for HCI and human-to-human interaction analysis. In HCI systems, multimodal signals can be used as input and/or output. Chapter 12 addresses multimodal inputs and focuses on the links between multimodality and cognition, namely the application of human cognitive processing models to improve our understanding of multimodal behaviour in different contexts, particularly in situations of high mental demand. The last section of this chapter discusses how patterns of multimodal input can inform and inspire the design of adaptive multimodal interfaces and the wide ranging benefits of such an adaptation. On the other hand, Chapter 13 presents examples of multimodal output, consisting of synthesised speech, facial motion and gesture, whereas Chapter 14 addresses the problem of representing multimodal databases in the context of the construction of interactive platforms for searching, browsing and navigating in multimedia collections. Finally, in Chapter 15, we present an application of multimodal signal processing for human-to-human interaction analysis, namely modelling interest in face-to-face conversations from multimodal nonverbal behaviour.

Part I

Signal Processing, Modelling and Related Mathematical Tools

Chapter 2

Statistical Machine Learning for HCI

Samy Bengio
Google, Mountain View, USA

2.1 INTRODUCTION

This chapter introduces the main concepts of statistical machine learning, as they are pivotal in most algorithms tailored for multimodal

Multimodal Signal Processing, ISBN: 9780123748256

signal processing. In particular, the chapter will cover a general introduction to machine learning and how it is used in classification, regression and density estimation. Following this introduction, two particularly well-known models will be presented, together with their associated learning algorithm: *support vector machines* (SVM), which are well known for classification tasks; and *hidden Markov models* (HMM), which are tailored for sequence processing tasks such as speech recognition.

2.2 INTRODUCTION TO STATISTICAL LEARNING

We introduce here briefly the formalism of statistical learning theory, which helps understanding of the general problem of learning a function from a set of training examples.

Let us assume that one has access to a training set of examples constructed as follows. Let $Z_1, Z_2, \ldots, Z_n$ be an n-tuple random sample of an unknown distribution of density $p(\mathbf{z})$. All Z_i are assumed to be independently and identically distributed (i.i.d.). Let D_n be a particular instance $= \{\mathbf{z}_1, \mathbf{z}_2, \ldots, \mathbf{z}_n\}$. D_n can be seen as a training set of examples $\mathbf{z}_i$, which can be used to solve various kinds of problems in machine learning, including the three most known categories of learning problems: *classification*, *regression* and *density estimation*[1]. Applications of such learning problems are vast, including speech processing (speech recognition, speaker recognition, keyword spotting, etc.), image processing (object recognition, face recognition, character recognition, etc.), prediction (time series, finance, geostatistics, etc.) and many more.

2.2.1 Types of Problem

The goal of *classification* is to determine the class label of a given example, out of a finite set of classes, based on a description of the example. In the simplest two-class classification problem, an example Z can be represented by a pair (X, Y) as follows

$$Z = (X, Y) \in \mathbb{R}^d \times \{-1, 1\}$$

---

1. Naturally, many other types of problems exist, such as time series prediction or ranking, but they can often be cast in one of the three main categories.

where X is the random variable describing the example, represented by a d-dimensional vector, and Y is the class random variable. In that case, the objective is: given a new $\mathbf{x}$, estimate the class posterior probabilities $P(Y=1|X=\mathbf{x})$ and $P(Y=-1|X=\mathbf{x})$ and select the class yielding the highest estimate.

In *regression*, the goal is to estimate a real valued scalar given some input vector. In this case,

$$Z=(X,Y)\in\mathbb{R}^d\times\mathbb{R}$$

where now $Y\in\mathbb{R}$. The objective becomes: given a new $\mathbf{x}$, estimate $E[Y|X=\mathbf{x}]$, the expected value of Y given $\mathbf{x}$.

Finally, in *density estimation*, the problem is generally to estimate the probability that a given example is drawn from the same probability distribution as those of the training set. In that case,

$$Z\in\mathbb{R}^d$$

and the objective is: given a new $\mathbf{z}$, estimate $p(\mathbf{z})$.

2.2.2 Function Space

To solve a classification, regression or density estimation problem, we define learning as the process of searching for a good function f in a given function space $\mathcal{F}$. This function space is usually parameterised like $f(\cdot;\theta)\in\mathcal{F}$ and the exact form of $f(\cdot;\theta)$ depends on the problem to solve. For instance, in regression, one could span the set of linear functions in $\mathbf{x}$:

$$\hat{y}=f(\mathbf{x};\mathbf{w},b)=\mathbf{w}\cdot\mathbf{x}+b,$$

where $\hat{y}$ represents the expected outcome and $\theta=\{\mathbf{w}\in\mathbb{R}^d,b\in\mathbb{R}\}$ is the set of parameters to estimate; in classification, one could span the same set of functions, but add a sign function to take a decision:

$$\hat{y}=f(\mathbf{x};\mathbf{w},b)=\text{sign}(\mathbf{w}\cdot\mathbf{x}+b);$$

finally, for density estimation, one could span the set of all Gaussian probability distributions:

$$\hat{p}(\mathbf{z})=f(\mathbf{z};\boldsymbol{\mu},\boldsymbol{\Sigma})=\frac{1}{(2\pi)^{\frac{|\mathbf{z}|}{2}}\sqrt{|\boldsymbol{\Sigma}|}}\exp\left(-\frac{1}{2}(\mathbf{z}-\boldsymbol{\mu})^T\boldsymbol{\Sigma}^{-1}(\mathbf{z}-\boldsymbol{\mu})\right),$$

where $\hat{p}(\mathbf{z})$ is the expected density at $\mathbf{z}$ and $\theta = \{\mu, \Sigma\}$ with μ and Σ, respectively, the mean and covariance matrix of the Gaussian distribution.

2.2.3 Loss Functions

To select the best function in a given function space, one needs to define what *best* means. This is done by defining a *loss function L*: $\mathcal{Z} \times \mathcal{F}$ which takes a data point $\mathbf{z}$ and a function f and returns how *bad* function f is for data point $\mathbf{z}$. For instance, in regression, one could use the *squared error*:

$$L(\mathbf{z},f) = L((\mathbf{x}, y),f) = (f(\mathbf{x}) - y)^2$$

where y is the expected outcome and $f(\mathbf{x})$ is the output of the function to evaluate. A good f should be such that the squared error should be small for most data points $\mathbf{z}$. In classification, one could simply count the number of errors made by f, as follows:

$$L(\mathbf{z},f) = L((\mathbf{x}, y),f) = \begin{cases} 0 & \text{if } f(\mathbf{x}) = y \\ 1 & \text{otherwise,} \end{cases}$$

and for density estimation, one could use the negative log likelihood provided by the current estimate:

$$L(\mathbf{z},f) = -\log p(\mathbf{z}).$$

2.2.4 Expected Risk and Empirical Risk

We are now able to define properly the goal of learning as to find a function f, in a set of functions $\mathcal{F}$, such that the average loss obtained from any example $\mathbf{z}$ drawn i.i.d. from $p(\mathbf{z})$ is minimised. This average loss of a given $f \in \mathcal{F}$ is called the *expected risk*, and it is defined as

$$R(f) = E_Z[L(\mathbf{z},f)] = \int_Z L(\mathbf{z},f)p(\mathbf{z})d\mathbf{z}.$$

Using the *Induction Principle*, we would ideally like to select $f^\star$ such that

$$f^\star = \arg\min_{f \in \mathcal{F}} R(f), \tag{2.1}$$

but since $p(\mathbf{z})$ is unknown and we do not have access to all $L(\mathbf{z},f)$ for all values of $\mathbf{z}$, this cannot be estimated. Instead, we can minimise the *empirical risk*, computed on a provided *training set* of examples, as follows:

$$f^{\star} = \arg\min_{f \in \mathcal{F}} \hat{R}(f, D_n) = \arg\min_{f \in \mathcal{F}} \frac{1}{n} \sum_{i=1}^{n} L(\mathbf{z}_i, f). \qquad (2.2)$$

Vapnik's statistical learning theory [1] can then be used to analyse in which conditions the empirical risk minimisation procedure converges (or not) to the expected risk.

2.2.5   Statistical Learning Theory

An important property of a good learning procedure is its consistency, which studies how it behaves when the number n of examples in the training set grows to infinity. In particular, what can be said about the expected and empirical risks when n tends to infinity? Figure 2.1 shows that, as n grows, the expected risk of $f^{\star}$ estimated using (2.2) will go down while its underlying empirical risk will go up, and both tend to the same value as n tends to infinity.

But what can we expect for finite values of n? The so-called Vapnik–Chervonenkis dimension, or *capacity* $h_{\mathcal{F}}$ of a set of functions $\mathcal{F}$, can help answer this question. The capacity is a measure of the size (or complexity) of the set of functions $\mathcal{F}$. In particular, for classification, the capacity $h_{\mathcal{F}}$ is defined as the largest n such that there exists a set of examples D_n such that one can always find an $f \in \mathcal{F}$ which gives the correct answer for all examples in D_n, for any possible *labelling*.

Hence, for instance, the capacity of a set of linear functions ($y = \mathbf{w} \cdot \mathbf{x} + b$) in d dimensions is $d + 1$. In the general case, however, it is usually impossible to estimate analytically the capacity of any set of functions. Fortunately, it is often enough to be able to rank sets of functions with respect to their capacity, without the need to know their exact value. For instance, it is clear that the capacity of linear functions is lower than that of quadratic functions, which in turn is lower than that of cubic functions, etc.

Vapnik has shown that one can bound the difference between the expected and the empirical risks as a function of n and $h_{\mathcal{F}}$. Figure 2.2

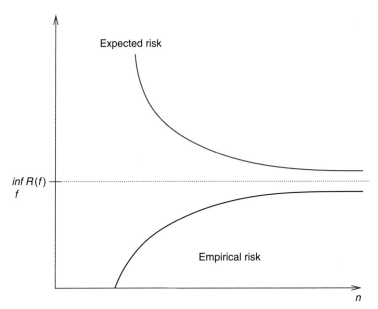

FIGURE 2.1   Consistency of the empirical risk minimisation procedure.

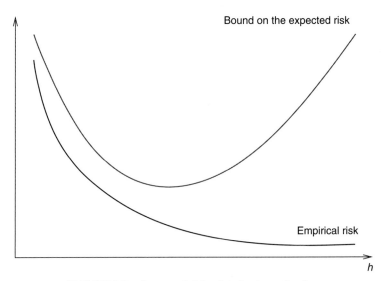

FIGURE 2.2   Structural risk minimisation – fixed n.

shows how both expected and empirical risks behave for a fixed number of training examples n, when varying the capacity $h_\mathcal{F}$. Hence, selecting the optimal set of function is a trade-off between having an expressive enough set of functions (large $h_\mathcal{F}$) such that it obtains a small empirical risk, and a constrained enough set of functions (small $h_\mathcal{F}$) such that it obtains a small expected risk.

In practice, most if not all machine learning models proposed in the literature have at least one *knob* that can be used to control the capacity, without explicitly computing it. For instance, the capacity of multi-layer perceptrons can be controlled by the number of hidden units and hidden layers; for SVMs, the capacity is controlled by the choice of the kernel; in HMMs, it is controlled by the number of states and the capacity of each emission distribution model; etc.

2.3 SUPPORT VECTOR MACHINES FOR BINARY CLASSIFICATION

The most well-known machine learning approach is the Support Vector Machine (SVM) [1]. It has been widely used in the last decade in various applications. Although its main successes are for classification problems [2, 3], extensions of the main approach to regression, density estimation, ranking and many others have been proposed. We describe in the following the simplest two-class classification framework.

Let us assume we are given a training set of n examples $T_{\text{train}} = \{(\mathbf{x}_i, y_i)\}_{i=1}^{n}$ where $\mathbf{x}_i \in \mathbb{R}^d$ is a d-dimensional input vector and $y_i \in \{-1, 1\}$ is the target class. The simplest binary classifier one can think of is the linear classifier, where we are looking for parameters ($\mathbf{w} \in \mathbb{R}^d, b \in \mathbb{R}$) such that

$$\hat{y}(\mathbf{x}) = \text{sign}(\mathbf{w} \cdot \mathbf{x} + b). \tag{2.3}$$

When the training set is said to be linearly separable, there is potentially an infinite number of solutions ($\mathbf{w} \in \mathbb{R}^d, b \in \mathbb{R}$) that satisfy (2.3). Hence, the SVM approach looks for the one that maximises the *margin* between the two classes, where the margin can be defined as the sum of the smallest distances between the separating hyper-plane and points of each class. This concept is illustrated in Figure 2.3.

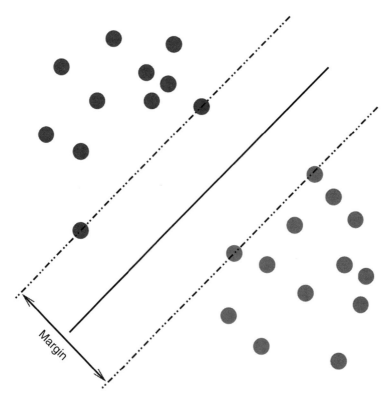

FIGURE 2.3   Illustration of the notion of margin.

It can be shown that maximising the margin is equivalent to minimising the norm of **w**, hence it can be expressed by the following optimisation problem:

$$\min_{\mathbf{w},b} \quad \frac{1}{2}\|\mathbf{w}\|^2 \tag{2.4}$$

$$\text{subject to } \forall i \quad y_i(\mathbf{w}\cdot\mathbf{x}_i+b)\geq 1$$

where the constraints express the classification problem to solve. To solve this problem, we usually start by writing the Lagrangian of (2.4), which introduces new variables α_i for each of the constraints in (2.4). Then instead of solving directly the problem, we solve its dual formulation, that computationally is more efficient:

$$\max_{\alpha} \quad \sum_{i=1}^{n} \alpha_i - \frac{1}{2} \sum_{i=1}^{n} \sum_{j=1}^{n} y_i y_j \alpha_i \alpha_j \mathbf{x}_i \cdot \mathbf{x}_j \qquad (2.5)$$

$$\text{subject to} \quad \begin{cases} \forall i \quad \alpha_i \geq 0 \\ \sum_{i=1}^{n} \alpha_i y_i = 0. \end{cases}$$

One problem with this formulation is that if the problem is not linearly separable, there might be no solution to it. Hence, one can relax the constraints by allowing errors with an additional hyper-parameter C that controls the trade-off between maximising the margin and minimising the number of training errors [4] as follows:

$$\min_{\mathbf{w},b} \quad \frac{1}{2} \|\mathbf{w}\|^2 + C \sum_{i} \xi_i \qquad (2.6)$$

$$\text{subject to} \quad \begin{cases} \forall i \quad y_i(\mathbf{w} \cdot \mathbf{x}_i + b) \geq 1 - \xi_i \\ \forall i \quad \xi_i \geq 0 \end{cases}$$

which dual becomes

$$\max_{\alpha} \quad \sum_{i=1}^{n} \alpha_i - \frac{1}{2} \sum_{i=1}^{n} \sum_{j=1}^{n} y_i y_j \alpha_i \alpha_j \mathbf{x}_i \cdot \mathbf{x}_j \qquad (2.7)$$

$$\text{subject to} \quad \begin{cases} \forall i \quad 0 \leq \alpha_i \leq C \\ \sum_{i=1}^{n} \alpha_i y_i = 0. \end{cases}$$

To look for non-linear solutions, one can easily replace $\mathbf{x}$ by some non-linear function $\boldsymbol{\phi}(\mathbf{x})$. It is interesting to note that $\mathbf{x}$ only appears in dot products in (2.7). It has thus been proposed to replace all occurrences of $\boldsymbol{\phi}(\mathbf{x}_i) \cdot \boldsymbol{\phi}(\mathbf{x}_j)$ by some kernel function $k(\mathbf{x}_i, \mathbf{x}_j)$. As long as $k(\cdot, \cdot)$ lives in a reproducing kernel Hilbert space (RKHS), one can guarantee that there exists some function $\boldsymbol{\phi}(\cdot)$ such that

$$k(\mathbf{x}_i, \mathbf{x}_j) = \boldsymbol{\phi}(\mathbf{x}_i) \cdot \boldsymbol{\phi}(\mathbf{x}_j).$$

Thus, even if $\boldsymbol{\phi}(\mathbf{x})$ projects $\mathbf{x}$ in a very high (possibly infinite) dimensional space, $k(\mathbf{x}_i, \mathbf{x}_j)$ can still be efficiently computed. Kernelised

SVMs can thus be solved as linear problems in a high-dimensional space that correspond to non-linear problems in the original feature space.

Problem (2.7) can be solved using off-the-shelf quadratic opti-misation tools. Note, however, that the underlying computational complexity is at least quadratic in the number of training examples, which can often be a serious limitation for most speech processing applications.

After solving (2.7), the resulting SVM solution takes the form of

$$\hat{y}(\mathbf{x}) = \text{sign}\left(\sum_{i=1}^{n} y_i \alpha_i k(\mathbf{x}_i, \mathbf{x}) + b\right) \tag{2.8}$$

where most α_i are zero except those corresponding to examples in the margin or misclassified, often called *support vectors* (hence the name of SVMs).

SVMs have been used in many successful applications and several off-the-shelf implementations are now readily available on the Web. The main limitation remains its training computational complexity, which grows at least quadratically with the number of examples, which hence limits the use of SVMs to problems with less than a few hundreds of thousands training examples. Furthermore, the size of the obtained model depends on the number of support vectors, which is expected to be linear with the number of training examples, hence it can become very inefficient at test time as well.

2.4 HIDDEN MARKOV MODELS FOR SPEECH RECOGNITION

HMMs [5] have been the preferred approach for many sequence processing applications, including most speech processing problems such as speech recognition, speaker verification, keyword detection and speaker segmentation, for the last 20 years or so. This section briefly describes the problem of speech recognition and shows how HMMs can be used for this problem.

Broadly speaking, an HMM can be seen as a special kind of *graphical model* [6], which studies the joint probability distribu-tions of several random variables. In the particular case where these random variables are organised as a time series, HMMs are one of the simplest, but still very powerful, models to estimate them.

2.4.1 Speech Recognition

Speech recognition is the most well-known successful application of HMMs; it is thus worth presenting the problem briefly before describing how HMMs are used in such a context. Note that it will be further developed in Chapter 3.

The starting point is a speech signal that is represented by a sequence of acoustic feature vectors, extracted by some front-end signal processor. Let us denote the sequence of acoustic feature vectors by $\bar{\mathbf{x}} = (\mathbf{x}_1, \mathbf{x}_2, \dots, \mathbf{x}_T)$, where $\mathbf{x}_t \in \mathbb{R}^d$. Each vector is a compact representation of the short-time spectrum. Typically, each vector covers a period of $10\,\mathrm{ms}$ and there are approximately $T = 300$ acoustic vectors in a 10 word utterance.

The spoken utterance corresponds to a sentence, which can be represented by a sequence of words $\bar{w} = (w_1, \dots, w_M)$. Each of the words belongs to a fixed and known vocabulary $\mathcal{V}$, that is, $w_i \in \mathcal{V}$.

The task of a speech recogniser is thus to predict the most probable word sequence $\bar{w}^\star$ given the acoustic signal $\bar{\mathbf{x}}$. Speech recognition is formulated as a *maximum a posteriori* (MAP) decoding problem as follows

$$\bar{w}^\star = \arg\max_{\bar{w}} P(\bar{w}|\bar{\mathbf{x}}) = \arg\max_{\bar{w}} \frac{p(\bar{\mathbf{x}}|\bar{w})P(\bar{w})}{p(\bar{\mathbf{x}})}, \qquad (2.9)$$

where we used Bayes' rule to decompose the posterior probability in the last equation. The term $p(\bar{\mathbf{x}}|\bar{w})$ is the probability of observing the acoustic vector sequence $\bar{\mathbf{x}}$ given a specified word sequence $\bar{w}$ and it is known as *the acoustic model*. The term $P(\bar{w})$ is the probability of observing a word sequence $\bar{w}$ and it is known as *the language model*. The term $p(\bar{\mathbf{x}})$ can be disregarded, because it is constant under the max operation.

2.4.2 Markovian Processes

Let q_t be a discrete event happening at time t, and let $q_1, q_2, \dots, q_T = q_1^T = \bar{q}$ be a sequence of such events. One can model the probability

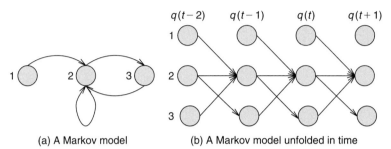

(a) A Markov model   (b) A Markov model unfolded in time

FIGURE 2.4   Two views of the same Markov process.

distribution of such a sequence as

$$P(\bar{q}) = P(q_1^T) = P(q_1)\prod_{t=2}^{T}P\left(q_t|q_1^{t-1}\right),\qquad(2.10)$$

where it is seen that the probability of event q_t depends on the values of previous events q_1^{t-1}. A first order Markovian assumption can simplify (2.10) as follows:

$$P\left(q_t|q_1^{t-1}\right) = P(q_t|q_{t-1})$$

which states that the probability of event q_t only depends on the value of event q_{t-1}.

Figure 2.4 shows two views of the same Markov process. On the left, a Markov process of three states is described, where each circle represents a state and each arrow represents the fact that a non-null probability exists to transit from a state to another at any given time. On the right, the same Markov process is shown, unfolded through time: each column represents a single time slice of the Markov process, showing all possible paths of q_t along time through the allowed transitions from column to column.

2.4.3 Hidden Markov Models

An HMM can be seen as a Markov process which event (often called *state*) is not observed, but from which one can observe a second random process (which thus depends on the hidden state). More formally, an HMM estimates $p(\bar{q}, \bar{\mathbf{x}}) = p\left(q_1^T, \mathbf{x}_1^T\right)$, the joint probability distribution of a discrete but hidden/unknown state variable q and a discrete or continuous but observable variable $\mathbf{x}^2$. An HMM can thus be decomposed as follows:

- a finite number of states N;
- a set of *transition probabilities* between states, which depend only on the previous state: $P(q_t|q_{t-1})$;
- a set of *emission densities*, which depend only on the current state: $p(\mathbf{x}_t|q_t)$ (where $\mathbf{x}_t \in \mathbb{R}^d$ is observed);
- an *initial state probability* distribution: $P(q_0)$.

Hence, using some additional Markovian assumptions, we get

$$p\left(q_0^T, \mathbf{x}_1^T\right) = p(q_0) \prod_{t=1}^{T} p(\mathbf{x}_t|q_t) P(q_t|q_{t-1}). \qquad (2.11)$$

Figure 2.5 shows two views of the same HMM. On the left, an HMM with three hidden states is described, where each numbered circle represents a hidden state, and each continuous arrow represents the fact that a non-null probability exists to transit from a state to another at any given time. Each non-numbered circle represents the

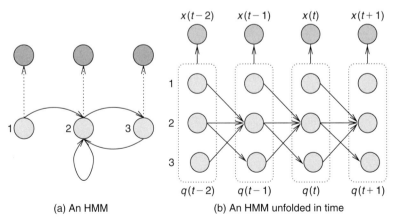

(a) An HMM   (b) An HMM unfolded in time

FIGURE 2.5   Two views of the same HMM.

2. In the following, we will mainly consider the continuous case where $\mathbf{x} \in \mathbb{R}^d$.

distribution of the observed variable for the corresponding hidden state. On the right, the same HMM is shown, unfolded through time: each column represents a single time slice, showing the underlying hidden Markov process and all possible paths of q_t along time through the allowed transitions from column to column.

Rabiner and Juang [5] describe three fundamental problems that need to be solved regarding HMMs:

- given an HMM parameterised by θ, how can we compute the *likelihood* of a sequence $p(\mathbf{x}_1^T; \theta)$;
- given an HMM parameterised by θ and a set of sequences $\bar{\mathbf{x}}_i \in D_n$, how can we *select the parameters* θ^* such that:

$$\theta^* = \arg\max_\theta \prod_{i=1}^{n} p(\bar{\mathbf{x}}_i; \theta);$$

- given an HMM parameterised by θ, how can we compute the *optimal path* $\bar{q}^\star$ through the state space given a sequence $\bar{\mathbf{x}}$:

$$\bar{q}^\star = \arg\max_{\bar{q}} p(\bar{\mathbf{x}}, \bar{q}; \theta)$$

2.4.4 Inference and Learning with HMMs

The first fundamental problem with HMMs regards inference, which means computing the likelihood of a sequence given an HMM model. This can be done by introducing the so-called forward variable $\alpha(i, t)$, which represents the probability of having generated the sequence $\mathbf{x}_1^t$ and being in state i at time t:

$$\alpha(i, t) \overset{\text{def}}{=} p(\mathbf{x}_1^t, q_t = i) \tag{2.12}$$

which can be shown to be efficiently estimated using the following recursive solution:

$$\alpha(i, t) = p(\mathbf{x}_t | q_t = i) \sum_{j=1}^{N} P(q_t = i | q_{t-1} = j) \alpha(j, t-1) \tag{2.13}$$

with the following initial condition

$$\alpha(i, 0) = P(q_0 = i) \rightarrow \text{ prior probabilities of each state } i. \tag{2.14}$$

This can then be used to compute the likelihood of a sequence as follows:

$$p(\mathbf{x}_1^T) = \sum_{i=1}^{N} p(\mathbf{x}_1^T, q_T = i) = \sum_{i=1}^{N} \alpha(i, T). \qquad (2.15)$$

A similar but *backward* variable $\beta(i, t)$ can also be defined as the probability to generate the rest of the sequence $\mathbf{x}_{t+1}^T$ given that we are in state i at time t:

$$\beta(i, t) \stackrel{\text{def}}{=} p(\mathbf{x}_{t+1}^T | q_t = i) \qquad (2.16)$$

which can be estimated by the following recursive equation:

$$\beta(i, t) = \sum_{j=1}^{N} p(\mathbf{x}_{t+1} | q_{t+1} = j)\beta(j, t+1)P(q_{t+1} = j | q_t = i). \qquad (2.17)$$

The learning problem consists of finding the best parameters θ of an HMM such that it maximises the likelihood of a set of observed sequences. This can be solved by the Expectation-Maximisation algorithm [7], and its particular implementation for HMMs, also known as the *forward-backward* or *Baum–Welsh* algorithm [8]. It yields an iterative but non-convex algorithm that provides parameter update equations that are based on the estimation of the transition and emission posterior probabilities $P(q_t = i, q_{t-1} = j | \mathbf{x}_1^T)$ and $P(q_t = i | \mathbf{x}_1^T)$, which in turn can be shown to be derived from the forward α and backward β variables for each state i and time step t of all training sequences. Although the state space of the HMM is exponential, each iteration of the Baum–Welsh algorithm can update its parameters with a computational complexity of $\mathcal{O}(N^2 \cdot T)$ per training sequence of size T.

The final problem consists of finding the optimal state sequence given an HMM and an observed sequence. This can be done efficiently thanks to the so-called Viterbi decoding algorithm [9], which follows basically the same derivation as for the forward variable α but where the operator $\sum$ has been replaced by a max instead:

$$V(i, t) = p(\mathbf{x}_t | q_t = i) \max_{j} P(q_t = i | q_{t-1} = j)V(j, t-1) \qquad (2.18)$$

where one also needs to keep for each max operator, the correspond-ing arg max state. One can then recover the best sequence of states by starting from the last state $i = \arg\max_j V(j, T)$ and applying the arg max operator recursively backward through the whole sequence.

2.4.5 HMMs for Speech Recognition

In the context of speech recognition, HMMs are used to estimate the acoustic model $p(\bar{\mathbf{x}}|\bar{w})$. Typically, word sequences $\bar{w}$ are broken down into sequences of smaller units, such as *phonemes*; a separate HMM is then trained for each phoneme model. Afterward, phoneme HMMs can be *concatenated* to create *word models* or even *sentence models*, which can then be used to estimate $p(\bar{\mathbf{x}}|\bar{w})$.

Finally, language models are used to estimate the probability of a given sequence of words, $P(\bar{w})$. It is often estimated by n-grams [10], where the probability of a sequence of M words $(\bar{w}_1, \bar{w}_2, \ldots, \bar{w}_M)$ is estimated as follows:

$$p(\bar{w}) \approx \prod_t p(w_t | w_{t-1}, w_{t-2}, \ldots, w_{t-M}) \qquad (2.19)$$

where each term of this product can be estimated on a large corpus of written document by simply counting the occurrences of each n-gram. Various smoothing and back-off strategies have been developed in the case of large n where most n-grams would be poorly estimated even using very large text corpora.

2.5 CONCLUSION

In this chapter, we have presented the main concepts of statistical machine learning, including capacity and expected risk, and we have seen how they apply to a variety of problems, such as classification, regression and density estimation. We have then presented two par-ticular machine learning models, the SVM, one of the most powerful classifiers recently proposed in the machine learning community, and the HMM, which is the most used model for sequential problems such as speech recognition. In both cases, we have introduced the mathematical derivations and presented some successes and limita-tions. Subsequent chapters will go into more details of these models when appropriate.

REFERENCES

1. V.N. Vapnik, The Nature of Statistical Learning Theory, second ed., Springer, 2000.
2. B. Boser, I. Guyon, V. Vapnik, A training algorithm for optimal margin classifiers, in: Fifth Annual Workshop on Computational Learning Theory, (1992) 144–152.
3. J. Shawe-Taylor, N. Cristianini, Support Vector Machines and Other Kernel-based Learning Methods, Cambridge University Press, 2000.
4. C. Cortes, V. Vapnik, Support-vector networks, Mach. Learn. 20 (3) (1995) 273–297.
5. L. Rabiner, B.-H. Juang, Fundamentals of Speech Recognition, first ed., Prentice Hall, 1993.
6. M.I. Jordan (Ed.), Learning in Graphical Models, MIT Press, 1999.
7. A. Dempster, N. Laird, D. Rubin, Maxmimum likelihood from incomplete data via the EM algorithm, J. R. Stat. Soc. Ser. B, 39 (1) (1997) 1–38.
8. L.E. Baum, T. Petrie, G. Soules, N. Weiss, A maximization technique occurring in the statistical analysis of probabilistic functions of Markov chains, Ann. Math. Stat. 41 (1) (1970) 164–171.
9. A.J. Viterbi, Error bounds for convolutional codes and an asymptotically optimum decoding algorithm, IEEE Trans. Info. Theory, 13 (2) (1967) 260–269.
10. C.D. Manning, H. Schutze, Foundations of Statistical Natural Language Processing, MIT Press, 1999.

Chapter 3

Speech Processing

Thierry Dutoit and Stéphane Dupont
Faculté Polytechnique de Mons, Belgium

Multimodal Signal Processing, ISBN: 9780123748256

3.1 INTRODUCTION

Text-to-speech (TTS) synthesis is often seen by engineers as an easy task compared with automatic speech recognition[1] (ASR). It is true, indeed, that it is easier to create a bad, first trial TTS system than to design a rudimentary speech recogniser. After all, recording numbers up to 60 and a few words ('it is now', 'A.M.', 'P.M.') and being able to play them back in a given order provides the basis of a working talking clock, while trying to recognise such simple words as 'yes' or 'no' immediately implies some more elaborate signal processing.

Users, however, are generally much more tolerant to ASR errors than they are willing to listen to unnatural speech. There is magic in a speech recogniser that transcribes continuous radio speech into text with a word accuracy as low as 50%; in contrast, even a perfectly intelligible speech synthesiser is only moderately tolerated by users if it delivers nothing else than 'robot voices', which cruelly lack naturalness.

In this chapter, we outline the main techniques that are available in today's ASR (Section 2.2) and TTS systems (Section 2.4) (including speaker recognition), with special emphasis on the concepts and on the requirements imposed by their implementation, as well as on the resulting limitations.

ASR is a major component in many spoken language systems. It enables the development of useful concepts for human–machine interfaces but also for computer-mediated human-to-human communication. Automatic transcription of massive amounts of audio and video data, to be used for information retrieval, is another application. This section will give an outline of where we stand in terms of technology, performance and applications.

ASR technology has been drawing from a range of disciplines, including digital signal processing, probability, estimation and information theories, and also, naturally, from studies about the production

---

1. In an international conference on speech processing, a famous scientist once brought up a tube of toothpaste (whose brand was precisely Signal^{TM}) and, pressing it in front of the audience, he coined the words: 'This is speech synthesis; speech recognition is the art of pushing the toothpaste back into the tube…'.

and perception of speech, and the structure of spoken language. Decades of research saw the emergence of a widely deployed architecture, making use of statistical modelling frameworks, including hidden Markov models (HMMs, Chapter 2, Section 2.4). Fundamentally novel paradigms are also being developed, but the flexibility of the traditional approach kept attracting research in the recent years, often focusing on improving or replacing individual components of the system, on addressing identified bottlenecks or on gradually relaxing assumptions made in these models.

Major technological progress is being recorded regularly. However, there are still technological barriers to flexible solutions and user satisfaction under all circumstances. These are likely related to weaknesses at all stages, including the possibly weak representation of grammatical and semantic knowledge compared to human knowledge. Current systems are also very sensitive with respect to environmental conditions. Background noise, crosstalk, transducer and channel properties are all known to severely affect their performance. A second aspect emphasised by current research is the sensitivity with respect to variation naturally present in speech. For instance, the lack of robustness to foreign accents precludes the use by specific populations. Many factors affecting the realisation of speech, such as regional, sociolinguistic or related to the environment or the speaker herself, create a wide range of variation that may not be modelled correctly (speaker, gender, speaking rate, vocal effort, regional accent, speaking style, nonstationarity and so on), especially when resources for system training are scarce.

A related area, speaker recognition, will be introduced too (Section 3.3). ASR and speaker recognition have developed connected research communities. They however have different fundamental challenges to achieve, on one side, invariance with respect to the speaker difference (the Holy Grail, robust speaker-independent (SI) speech recognition) and on the other side, invariance with respect to what is being said (text-independent speaker recognition). Links between both communities are however developing more recently, in particular regarding speaker adaptation technologies, to the benefit of both domains.

3.2 SPEECH RECOGNITION

Statistical modelling paradigms, including HMMs and their extensions, are key approaches to ASR. Using proper assumptions, these technologies provide a mean to factorise the different layers of the spoken language structure. Several major components hence appear. First, the speech signal is analysed using *feature extraction* algorithms. The *acoustic model* is then used to represent the knowledge necessary to recognise individual sounds involved in speech (phonemes or phonemes in context). Words can hence be built as sequences of those individual sounds. This is represented in a *pronunciation model*. Finally, the *language model* (LM) is being used to represent the knowledge regarding the grouping of words to build sentences. The structure of those models is generally guided by scientific knowledge about the structure of written and spoken language, but their parameters are estimated in a data-driven fashion using large speech and text corpora. At runtime, the key role of determining what is the sequence of words that best matches an input speech signal is taken by the *decoder*, which is basically a graph search making use of these different models. Search algorithms can be quite straightforward, for instance for recognising words spoken in isolation, but may become more complex for recognising very large vocabulary continuous speech. Besides, multichannel and multimodal techniques specifically developed for ASR have been proposed in the literature. They can rely on alternate acoustic sensors or alternate modalities, like lip contour information. In the framework of multimodal systems, measurement on how certain each modality is, using so-called confidence measures, can also be a useful component for multimodal fusion.

3.2.1 Feature Extraction

After sampling and quantisation (typically 16 kHz and 16 bits), the speech signal is known to still be relatively redundant. This property has in particular been exploited for transmitting speech through low bit rate channels, where speech coders/decoders (codecs) are designed to extract compact representations that are sufficient for high-quality (or at least intelligible) reconstructions of the signal at the back-end. In ASR systems, compact representations are also sought. Signal processing algorithms are used to extract salient feature vectors, maintaining the information necessary for recognising speech and discarding the remainder. This step is often called 'feature extraction'.

This step basically relies on the *source-filter* model in which speech is described as a source signal, representing the air flow at the vocal folds, passed through a time-varying filter, representing the effect of the vocal tract. Speech recognition essentially relies on the recognition of sequences of phonemes which are mostly dependent on vocal tract shapes. A central theme here is hence the separation of the filter and source parts of the model.

Intuitively, the general approach will be to extract some smooth representation of the signal power spectral density (characteristic of the filter frequency response), usually estimated over analysis frames of typically 20–30 ms fixed length. Such short analysis frames are implied by the time-varying nature of both the source and the filter. Several signal processing tools are often used in feature extraction implementations. These include the short-time Fourier transform allowing to obtain the power and phase spectra of short analysis frames. A second tool is Linear Predictive Coding (LPC) in which we model the vocal tract by an all-pole filter. Another tool is the *cepstrum*, computed as the inverse short-time Fourier transform of the logarithm of the power spectrum. It can be shown that low order elements of the cepstrum vectors provide a good approximation of the filter part of the model.

Knowledge about the human auditory system has also found its way in speech analysis methods. Some forms of models of the nonlinear frequency resolution and smoothing of the ear are regularly being used. This is the case for the Mel-Frequency Cepstrum Coefficients (MFCCs), as well as for the Perceptual Linear Prediction (PLP) techniques, where the cepstral coefficients are computed from a spectrum that has been warped along nonlinear spectral scales (mel and Bark scale, respectively). The PLP technique additionally relies on LPC for further smoothing. Other perceptual properties, such as temporal and frequency masking, are also being investigated, for instance to increase the robustness to background noise, a central issue in today's ASR systems. Another avenue has been to address the typical analysis assumptions, for instance through signal representations that can better handle the inherent nonstationarity in speech [1] or make use of information that is traditionally discarded, such as phase [2] or prosody. Combining a range of analysis schemes in an ad hoc fashion, or using proper statistical formalisms, is often proposed too. These approaches are obviously key when dealing with signals captured using multiple sensors in a multimodal framework for instance.

To finish, it should be noted that the speech representations described here are very sensitive to noise, both additive (car noise, air conditioning noise and so on) and convolutive (the fact that the signal is collected using a microphone and an analogue path that can exhibit specific responses), as well as to the diversity of voice timbre and quality.

3.2.2   Acoustic Modelling

HMMs (see Chapter 2, Section 2.4) are one of the fundamental approaches around which the ASR research community and most free toolkits and commercial products have been developed. Speech signals can indeed be seen as stochastic time-series that have a finite state structure, and HMMs are proper models for those.

The HMM approach draws from linguistic and phonetic sciences, and uttered sentences are considered to be built of sequences of words, themselves built using sequences of phonemes. Allophones or phonemes in context (triphones, quinphones and so on) are being used within context dependent (CD) modelling to better account for the coarticulation between neighbouring phonemes. These phonetic units are seen as the basic constituents of speech and form the state set of an HMM structure, i.e., stochastic processes are associated to these phonetic units. They are designed to model the distribution of feature frames introduced in the previous sections. The parameters of these models can be estimated over recorded training material, and several modelling approaches are possible. Using mixtures of multidimensional Gaussian distributions (Gaussian mixture models, GMMs) is probably the most widely used method and directly allows modelling the state processes as continuous parametric distributions. Minor modifications in the formalism broaden its use to other modelling schemes. For instance, vector quantisation (VQ) of the feature frames permits using discrete distributions. Discriminant approaches can also be developed, where the probability distribution is not explicitly modelled. One such approach uses support vector machines (SVMs) and various forms of artificial neural networks (ANNs) [3], such as multi-layer perceptrons or recurrent neural networks.

During operation, the approach consists in finding the sequence of words, and hence of HMM states, that most likely produces the observed time series of feature frames, given the model. The HMM is called hidden because, given its stochastic nature, one cannot know for sure the state sequence given a sequence of feature frames. Those states are called *hidden states*, whereas the feature frames are called the *observations*, or sometimes the observed states. Finding the word sequence is the role of search algorithms, often based on dynamic programming principles, including large vocabulary decoding approaches.

The HMM formalism is briefly reminded here, as it is useful to introduce other elements of the ASR architecture, such as pronunciation modelling, as well as language modelling.

From a sequence of observations O coming from the feature extraction step, the task of speech recognition consists in estimating the word sequence W, according to

$$\hat{W} = \arg\max_{W} P(W|O). \tag{3.1}$$

After applying the Bayes rule, we have the following:

$$\hat{W} = \arg\max_{W} p(O|W)P(W). \tag{3.2}$$

The second factor $P(W)$ will be handled by the LM (as shown in the following section). The first factor, $p(O|W)$ is the probability of observing O knowing the sequence of words W and is usually handled using HMMs. This goes as follows. The word sequence W is represented by a sequence of individual HMM states. However, there can be many different valid state sequences for a given sentence because some words may have multiple possible pronunciations and also because the temporal boundaries between states are allowed to vary. Hence, we can write $p(O|W)$ as a sum of probabilities for all possible state sequences S:

$$p(O|W) = \sum_{\text{all } S} p(S|W)P(O|S,W). \qquad (3.3)$$

For an observation sequence $O = (o_1, o_2, \ldots, o_T)$ containing T observations (typically, observations are feature vectors) and a particular state sequence $S = (s_1, s_2, \ldots, s_T)$ (of length T too), we can hence rewrite the first factor (dropping W for convenience), by applying the *first-order* (Markov) *assumption*:

$$P(S) = P(s_1) \prod_{t=2}^{T} P(s_t|s_{t-1}). \qquad (3.4)$$

This factor involves the so-called HMM *transition probabilities*. For the second factor, we apply the *output-independence assumption*, which states that the observations are independent and identically distributed and that their probability distribution only depends on the current state at time t. We hence have

$$P(O|S) = \prod_{t=1}^{T} p(o_t|s_t). \qquad (3.5)$$

This expression involves the so-called HMM *emission probabilities*, handled by the acoustic model, which can rely on various frameworks, including GMMs as introduced previously.

The ASR literature is very rich in publications focusing on experiments around the model topology, or the nature of the states, on methodologies to tie parameters across states and on modelling approaches, just to name a few. As for feature extraction, a significant part of recent research has also attempted to overcome the typical assumptions made in the HMM framework, such as the first-order and output independence assumptions. As a matter of fact, HMMs can also be seen as a particular case of dynamic Bayesian network (DBN), which are also being explored for speech recognition [4], as they provide more room for investigating modelling variants.

The HMM and acoustic modelling components can also be the proper level to tackle the detrimental effect of background noise. More importantly, it is also the appropriate level to address effects due to the inherent (intrinsic) variability of speech, such as speaker specificities. In this case, *acoustic model adaptation techniques* are crucial. They are based on training algorithms allowing to tune the model parameters to specific acoustic conditions, or voices, using a reasonably small amount of adaptation samples. Maximum likelihood linear regression (MLLR) [5] is a good example, which has been applied to both speaker variability and noise. The idea is to estimate an affine transformation of the acoustic model parameters, by maximising the likelihood of adaptation data from a particular speaker, to shift the parameters of a generic SI acoustic model closer to those of a speaker-dependent acoustic model. Several other techniques have been proposed, with a recent focus on 'fast' adaptation, i.e., adaptation with few data [6].

3.2.3   Language Modelling

As shown in expression (3.1), the general ASR formalism requires the LM to provide an estimation of the probability of word sequences. To put it differently, this allows us to introduce other sources of knowledge.

LMs vary drastically depending on the application. On one side, limited scope applications make it possible to define exhaustively the admitted language as a finite set of words or sentences. This is the case for simple command and control tasks. On the other side, there are applications where no particular constraints are imposed on what the user can say. The typical example is automatic real-time dictation systems or transcription systems used to index audio or video content. In these cases, a probabilistic approach is often taken through the use of *n-gram* LMs.

A *n*-gram is actually a Markov chain. Formally, $P(W)$ can be decomposed as

$$
\begin{aligned}
P(W) &= P(w_1, w_2, \ldots, w_M) \\
&= P(w_1)P(w_2|w_1) \ldots P(w_M|w_1, w_2, \ldots, w_{M-1}) \qquad (3.6) \\
&= \prod_{m=1}^{M} P(w_m|w_1, w_2, \ldots, w_{m-1}),
\end{aligned}
$$

where $P(w_m|w_1, \ldots, w_{m-1})$ is the probability that word w_m will follow the word sequence $w_1, \ldots, w_{m-1}$ that appeared previously. Using the full word history, as expressed here, would be intractable. This is why n-grams assume that the terms in the product only depend on the $n-1$ preceding words. N-grams typically have orders ranging from two (unigrams) to four (trigrams) or even four. For a trigram, for instance, we would have $P(w_m|w_1, \ldots, w_{m-1}) = P(w_m|w_{m-2}, w_{m-1})$.

This is nevertheless just the tip of the iceberg regarding language modelling. Other forms of LMs have been proposed, including class-based or multi-class LMs [7], where the probabilities depend on classes to which the word belongs rather than on words themselves. Also, an important issue with LMs is related to the inherent sparsity of training data, especially when one tries to make use of high order n-grams. A range of 'smoothing' approaches have been proposed to produce more robust estimates for previously unseen word sequences. Just like for acoustic modelling, which works better when trained for a specific speaker, LMs designed for a specific application domain will have a better predictive power. Domain specific data are often very scarce however. LM adaptation techniques have thus been the subject of some research, including cache-based or topic-adaptive LMs. A complete account on language modelling, including the many practical issues not outlined here, can be found in [8].

3.2.4 Decoding

Continuous speech recognition is a search problem which relies on acoustic, pronunciation and LMs. The modules implementing these search algorithms are often called *decoders*. Basically, the search consists in finding the path (sequence of HMM states) that best matches the audio signal, as expressed in equation (3.1).

The Viterbi algorithm, making use of the concept of dynamic programming, can be applied here, understanding that the best search path is composed of subpaths that are optimal themselves. This is straightforward for isolated word recognition. For continuous speech recognition however, the LMs impose some particular constraints that need to be considered and the complexity of the search problem is essentially related to the complexity of the LMs, such as the

n-gram model introduced earlier. Sticking to order one (bigram LM), the number of states in the search space is proportional to the vocabulary size. For higher orders (n-grams), the search space becomes more complex. With a trigram, it is proportional to the square of the number of words.

Several strategies have been designed to speed up the search. A first one is to represent the lexicon as a tree, where the words are prefixes for other words, leading to a significant reduction of the search space. Beam search strategies are also used to discard search paths that are less likely to lead to the desired outcome. Finally, multiple-pass decoding allows to gradually introduce more complex knowledge sources. In this case, each pass is required to generate a list of best state sequences or a lattice representing this list. More complex LMs can hence be applied on top of these reduced search spaces. These advances, together with advances in hardware price-performance ratio, actually made very large vocabulary real-time speech recognition a reality, even on pocket devices.

3.2.5   Multiple Sensors

Human communication by essence involves several modalities. On one part, verbal aspects generate a range of 'alternate' signals related to the speech production process: vocal folds vibrations, air flow and also, essentially, movements of the articulators such as the jaws, tongue, cheeks and lips. Nonverbal aspects are also implied, including body motion, communicative gestures with the hands and the head and facial expressions. Multimodality involving emotions, gestures and other modalities is out of the scope of this chapter (see Chapter 7 for an overview of multimodal analysis). We will hence be sticking to speech here. Interesting research results have been published in the use of alternate sensors able to capture other correlates of vocal expression. Most of these studies are drawing from the inherent immunity of alternate sensors to background acoustic noise, a very detrimental factor for ASR performance.

Various authors try to make use of noise-immune sensors attached to the body. In addition to conduction through the air, speech is indeed naturally conducted within the body tissues and up to the skin, bones and ear canal. In [9], a so-called nonaudible murmur (NAM) microphone is used just by itself as an alternative to an ordinary microphone. The NAM device is based on a stethoscopic system and is attached to the skin of the user, for instance behind the ear. The work focuses on one major issue in using alternative devices: the need to estimate an HMM model that is valid for the new signals. The paper proposes to use an adaptation procedure based on MLLR. MLLR adaptation is also used in [10]. Here, the sensor is an inner microphone that captures the speech signal behind an acoustic seal (or earplug) in the auditory canal. The ASR performance and robustness based on the inner microphone and a close-talk microphone are compared. It is shown on data recorded in real noisy environments that the inner microphone produces the same performance as the standard microphone at 15–20 dB higher noise level. In [11], the combined use of a standard and a throat microphone is investigated. The combination technique is based on piecewise-linear transformation of a feature space to another. The source feature space is defined by a feature vector that contains both standard and throat microphone features in noisy conditions, and the target space is defined by the clean speech features. In [12], the sensor is a bone-microphone coupled with a regular close-talk microphone. The device has the look and feel of a regular headset, but a bone-microphone component is applied just above the zygomatic bone. Here, the proposed technique also consists in estimating the clean speech features from the features computed from both microphones coupled with a using a mapping technique from the bone-microphone features to the clean speech features. In [13], a throat-microphone has also been used in combination with a close-talk microphone. It is shown that simply using the alternate sensors for detecting nonspeech regions already brings a large improvement in terms of robustness.

These papers illustrate the variety of devices that can be used as a replacement or a complement to regular microphones. Sensors that are not relying on acoustic signals can also be considered. For instance, radar-like technology applied around the vocal tract and electromyography applied to the muscles of the jaws have also been proposed in the literature for improved speech enhancement or recognition.

Another category of approaches is to make use of moving images of the speaker's lips. The visual modality is known to be very useful for hearing impaired people. It provides information about the speaker localisation and contains complementary information about the place of articulation of phonemes, thanks to the visibility of the tongue, teeth and lips. Approaches described in the literature [14] complement the audio feature extraction by a visual feature extraction that extract characteristic vectors of the articulators contour and appearance. Depending on the approach, the integration of both feature stream can then happen more or less early in the recognition chain. The multimodal integration methods for audio-visual speech recognition are described in more detail in Chapter 9.

The results reported in the literature illustrate how ASR can benefit from alternative and complementary sensing devices. In [10] for instance, for a small vocabulary isolated words recognition task, using the alternate acoustic sensor placed in the auditory canal gives the same ASR error rate than using a 'traditional' close-talk microphone, but for background noise levels that can range from 5 to 15 dB above. This is a result of the stronger immunity of the proposed sensor. Audio-visual approaches too can result in important signal-to-noise ratio gains, around 7 dB across small and large vocabulary tasks, as reported in [14]. Significant gains are also reported in clean conditions, as well as for both normal and impaired speech. This is a consequence of the complementarity of acoustic and facial cues. These approaches could possibly handle nonstationary noises too, or even background speech, for which current speech enhancement or model adaptation techniques still have a hard time to deal with.

3.2.6   Confidence Measures

Earlier we mentioned the possibility for search algorithms to output lists containing the N best word sequences. This approach is often used in multimodal recognition relying on the *late-integration approach*, where a decision is to be made using N-best lists from different modalities (see Chapter 8 for more details and also for other integration methods).

Confidence measures [15, 16] are a complementary tool for multimodal systems. These are designed to quantify how reliable the recognised sequence is. They are used to reject out-of-vocabulary words or uncertain recognition due to the background noise in the audio or else to perform word spotting (recognition of a limited number of keywords in an audio stream, generally without using any LM). These are obviously important concepts when the ASR is considered as a component in a larger multimodal dialog system, where the outcomes from different modalities can be better integrated using such confidence measures. Confidence measures are also extremely useful in dialogue systems, potentially guiding it towards a less frustrating experience, especially when there is a need to redirect the caller to a human operator.

3.2.7 Robustness

Current ASR systems show their limits when dealing with noisy environments [17]. A range of approaches for improving their robustness has been extensively discussed in the literature. Speech enhancement, often relying on Bayesian filtering techniques, is very common. When they are designed to estimate a clean speech signal, they are also useful in the framework of person-to-person communication systems. Instead of trying to provide the ASR module with a cleaner input signal, one can also 'contaminate' the training database so that the estimated acoustic model parameters are more appropriate for recognition in noisy conditions. Processing applied on the feature streams has also been proposed, such as the filtering of the trajectories of the feature vector elements. Finally, techniques that consider the most degraded parts of the signal as missing data have also been proposed.

Inter- and intra-speaker variability can also be considered as some kind of noise for the ASR system. It is well known indeed that speech not only conveys the linguistic information (the message) but also conveys a large amount of information about the speaker himself: gender, age, social and regional origin, health and emotional state and, with a rather strong reliability, his identity. In addition to intraspeaker variability (emotion, health, age), it is also commonly admitted that the speaker uniqueness results from a complex combination of physiological and cultural aspects [18, 19]. Current ASR systems are still much less efficient than humans when dealing with these variation sources.

Characterising the effects of such variations, together with design techniques to improve ASR robustness, is a major research topic [20]. Speech nonstationarity is a first obvious theme. The effect of coarticulation has motivated studies on segment-based [21], articulatory and CD modelling techniques. Even in carefully articulated speech, the production of a particular phoneme results from a continuous gesture of the articulators, coming from the configuration of the previous phonemes and going to the configuration of the following phonemes (coarticulation effects may indeed stretch over more than one phoneme). In different and more relaxed speaking styles, stronger pronunciation effects may appear and often lead to reduced articulation, some of these being particular to a language (and mostly unconscious). Others are related to regional origin and are referred to as accents (or dialects for the linguistic counterpart) or else to social groups and are referred to as sociolects. Although some of these phenomena may be modelled implicitly by CD modelling techniques [22], their impact may be more simply characterised at the pronunciation model level. At this stage, phonological knowledge is helpful [23], especially in the case of strong effects like foreign accents. Fully data-driven techniques have also been proposed though [24, 25]. Speaker-related (as well as gender-related) spectral characteristics have been identified as another major dimension of speech variability. Specific models of frequency warping (based on vocal tract length differences) have been proposed [26, 27], as well as more general feature compensation and model adaptation techniques. Speech also varies with age, due to both generational and morphological reasons. The two 'extremes' of the range are generally put at a disadvantage [28] due to the fact that available corpora used for model estimation are typically not designed to be representative of children and elderly speech. Again, some general adaptation techniques can however be applied to counteract this problem.

Intraspeaker variation sources are admittedly affecting the signal and ASR systems too. A person can change his voice to be louder, quieter, more tense or softer or even a whisper. Also, some reflex effects exist, such as speaking louder when the environment is noisy, known as the Lombard reflex. Speaking fast or slow also has influence on the speech signal [29]. This affects both temporal and spectral characteristics, both affecting the acoustic models. Obviously, faster speaking rates also result in more frequent and stronger pronunciation changes. Over-articulation is the opposite case and causes some trouble too. It's been reported that when users are speaking slower and emphasise all syllables, as when making a slow dictation, they notice that the ASR system does not recognise their words properly. This only increases the problem, as acoustic models are often being trained using 'regular' speech. Emotions are also becoming a hot topic, as they can indeed have a negative effect on ASR and also because added value can emerge from applications that are able to identify the user emotional state (frustration due to poor usability for instance) [30].

Regarding those aspects, extensions of the HMM formalism, using DBNs [31], or where variability gives rise to additional hidden variables [32], together with the availability of larger and more diverse corpora for acoustic model and pronunciation model studies, hold future promises.

3.3 SPEAKER RECOGNITION

3.3.1 Overview

While speech recognition deals with recognizing what is being said, speaker recognition tends to deal with who is speaking. This discipline also has a multi-decades long history, fueled by perspectives of applications as an easy biometric approach for authentication in the forensic and security fields, as well as applications for automatically enriching the transcription of broadcasted spoken content with speaker identity.

Speaker recognition technology attempts to draw from the speech correlates of person morphology (like the size and properties of the folds and vocal tract), as well as from speaker specific speaking styles, including pronunciation, as well as prosody, the latter being most of the time disregarded in the companion ASR technology. Speaker recognition covers two kinds of methodologies: speaker verification (also known as authentication), and speaker identification. Speaker verification [33a] consists in verifying whether a user claimed identity is correct, for instance acting as a 'voice print' verification to be used for providing access to a secure system. Speaker identification consists in recognizing a speaker identity, amongst a set of N known identities, for instance to be used for identifying who is speaking in a conversation.

Whereas ASR can be developed without knowledge of the user voice (speaker-independent systems), speaker recognition requires a prior enrolment phase, where voice is recorded and used to estimate the parameters of a model specific to the speaker (voice print). Technically, the user does not need to be aware of this enrolment phase however. During operation, an incoming voice is analyzed and compared to one or several enrolled voices, depending on the scenario. For identification, the speaker for which the model best matches the input utterance is selected. For verification, the speaker is accepted if its model matches the input utterance well enough. Also, several operation methods exist. In the text-dependent method, the user is asked to produce a given sentence, the same than the one produced during enrollment. In the text-independent method, any spoken input will do. Finally, text-prompted methods also exist, where the systems asks the user to pronounce a given utterance, which can be changed every time.

Speaker recognition research and development has heavily been sharing with ASR R&D and involves a similar set of tools, including feature extraction, HMMs, VQ, statistical modelling and classification techniques such as GMMs, SVMs and ANNs.

Speaker recognition also starts by a feature extraction module, with similar features than the one used for ASR, for instance MFCCs and their temporal derivatives. Other kinds of feature extraction schemes that are not tremendously useful for ASR have been successfully applied to speaker recognition. This include supra-segmental features like prosodic cues [33b] and features representative of speaking styles and sociolect, for instance using probabilities of particular phoneme or word sequences [34]. The speaker-specific modelling that follows feature extraction is very similar to the acoustic modelling approaches used in ASR, except that one model is required per speaker, whereas most ASR systems are equipped with only a single SI model. One popular approach is indeed to use ergodic HMMs where individual states represent phoneme categories and are modelled using GMMs. These approaches work well when enough enrolment data are available.

Variants of this approach exist and can differ in the way the phoneme categories are recognised and the way the speaker-specific models are obtained. More generally, the use of ASR approaches has increasingly been considered in speaker recognition systems. For instance, instead of implicitly recognising the phoneme categories using the speaker-specific model, a SI acoustic model can also be used to first recognise those categories, and the speaker-specific model is then used for those categories to match the input utterance.[2] Regarding the speaker specific models, approaches where these are obtained from an SI model (sometimes known as a background model), using speaker adaptation techniques, such as maximum a posteriori or MLLR adaptation, have also been proposed [35]. When enrolment data are scarcer, fast speaker adaptation techniques, such as speaker clustering and eigenvoices, may be used too. Experimental results [36] show that conventional GMMs perform better if there are abundant enrolment data, whereas fast speaker adaptation GMMs using eigenvoices perform better if enrolment data are sparse. Also, 'by-products' of speaker adaptation techniques can be put to use. For instance, speaker adaptation parameters, such as MLLR matrices, have been proposed as features vectors for speaker verification [37].

Alternate approaches using classification techniques (rather than generative modelling) have been proposed, for instance using two-class classification using SVMs [38], where the classification is between a particular speaker of interest and the background (a set of speakers covering the entire population). As in ASR, speaker identification and verification approaches can be used stand-alone, or in combination, often yielding significant performance gains.

Formally, for speaker verification, the typical baseline approach hence goes like this and can be formulated as a hypothesis test. We have a sequence of observation (feature) vectors O and want to test between hypothesis H_0 *(O has been generated by claimed speaker S)*

2. Note that this applies to the text-independent methods only, as in the text-dependent and text-prompted methods, the sequence of pronounced phonemes is normally known.

and hypothesis H_1 *(O has not been generated by speaker S)*. The likelihood ratio test can hence be used:

$$\frac{p(O|H_0)}{p(O|H_1)} \begin{cases} > \theta & \text{accept } H_0 \\ < \theta & \text{accept } H_1 \end{cases}, \tag{3.7}$$

where $p(O|H_0)$ and $p(O|H_1)$ can be estimated using HMMs and GMMs, as stated before.

As we said, speaker verification systems accept a user if the model of the claimed user matches the input utterance well enough. This implies the use of a threshold [θ in equation (3.7)]. If this threshold is set too low, impostors will more likely be accepted (false acceptance), whereas if it is set too high, valid user will more likely be rejected (false rejection). Hence, false acceptance and false rejection form a characteristic performance profile for a particular system, with performance metrics coming from detection and binary classification theory (see Chapter 2). The performance of a verification system indeed uses different evaluation metrics than speech recognition and speaker identification systems, and its performance is normally independent on the size of the test population (whereas the performance of an identification system will decrease as the user set size increases). Receiver operating characteristics displaying the correct acceptance rate versus the false acceptance rate, as well as equal error rate (EER) corresponding to the point in the characteristic where the false acceptance rate is equal to the false rejection rate, are commonly used. EERs between 1 and 5% are often reported.

3.3.2   Robustness

ASR and speaker recognition share similar challenges regarding their robustness. Speaker recognition systems indeed show some of their limits when dealing with a changing communication channel, for instance when using the system from a different telephone than the one used during the enrolment phase or with noisy speech. Techniques similar to the ones applied in robust ASR remain valid [39]. Also, dealing with intraspeaker variation is an additional difficulty. Voice changes over time, due to aging or health conditions, stress the speaker recognition system.

The tuning of the decision threshold (θ) for a particular application is also very troublesome as the scores involved in the likelihood ratios calculation can vary according to speaker and environmental changes. Score normalisation techniques have hence been introduced explicitly to cope with score variability and to allow an easier tuning of the decision threshold. Several techniques are described in [34].

3.4 TEXT-TO-SPEECH SYNTHESIS

Delivering *intelligibility* and *naturalness* has been the Holy Grail of speech synthesis research for the past 30 years. Speech *expressivity* is now increasingly considered as an additional objective to reach. Add to it that engineering costs (computational cost, memory cost, design cost for having another synthetic voice or another language) have always had to be taken into account, and you will start having an approximate picture of the challenges underlying TTS synthesis.

Although several paths have been and are still tried to reach these goals, we will concentrate here on the ones which have currently found their way to commercial developments, namely *concatenative synthesis based on a fixed inventory*, *concatenative synthesis based on unit selection* and *statistical parametric synthesis*. Other techniques (among which *rule-based synthesis* and *articulatory synthesis*) are handled in more general textbooks (such as the recent book by Taylor [40]). But since a TTS synthesis system requires some front end analysis, we first start with a short description of the natural language processing (NLP) aspects of the problem. Nevertheless, these concepts are fully covered in Chapter 4.

3.4.1 Natural Language Processing for Speech Synthesis

The NLP module of a TTS system produces a phonetic transcription of the input text, together with some prediction of the related intonation and rhythm (often termed as prosody); the DSP module transforms this symbolic information into speech.

A preprocessing (or *text normalisation*) module is necessary as a front-end because TTS systems should in principle be able to read any text, including numbers, abbreviations, acronyms and idiomatics, in any format. The preprocessor also performs the (not so easy) task of finding the end of sentences in the input text. It organises the input sentences into manageable lists of word-like units and stores them in the internal data structure. The NLP module also includes a morpho-syntactic analyser, which takes care of part-of-speech tagging and organises the input sentence into syntactically-related groups of words. A phonetiser and a prosody generator provide the sequence of phonemes to be pronounced as well as their duration and intonation. Once phonemes and prosody have been computed, the speech signal synthesiser is in charge of producing speech samples which, when played via a digital-to-analogue converter, will hopefully be understood and, if possible, mistaken for real, human speech.

Although none of these steps is straightforward, the most tedious one certainly relates to prosody generation. *Prosody* refers to properties of the speech signal which are related to audible changes in pitch, loudness, syllabic length, and voice quality. Its most important function is to create a segmentation of the speech chain into groups of syllables, termed as *prosodic phrases*. A first and important problem for a TTS system is then to be able to produce natural sounding intonation and rhythm, without having access to syntax or semantics. This is even made worse by the important sensitivity of the human ear to prosodic modifications of speech (more specifically, those related to pitch): even slightly changing the shape of a pitch curve on a natural syllable can very quickly lead to a signal which will be perceived as artificial speech.

In modern TTS systems, prosody is only predicted on the symbolic level, in the form of *tones,* based on some linguistic formalism of intonation. Tones are associated to syllables and can be seen as phonological (i.e., meaningful) abstractions which account for the acoustic realisation of intonation and rhythm. A tonetic transcription prediction typically assigns 'high' (H) and 'low' (L) tones to syllables, as well as stress levels, and possibly organises them into prosodic groups of syllables. This related theory has been more deeply formalised into the *Tones and Break Indices* transcription system [41]. A still more recent trend is to avoid predicting tones. In this case, contextual morpho-syntactic information, together with some form of prediction of prosodic groups, is directly used to find speech synthesis units in a similar context.

3.4.2 Concatenative Synthesis with a Fixed Inventory

Producing speech samples automatically does not merely reduce to the playback of a sequence of prerecorded words or phonemes, due to *coarticulation*, caused by the inertia of the articulatory system. Thus, producing intelligible speech requires the ability to produce continuous, coarticulated speech. More precisely, transients in speech are more important for intelligibility than stable segments, while modifying stable segments (e.g., the centre of vowels) can very easily affect naturalness. Speech, indeed, is never really periodic nor stable. Even sustained vowels exhibit small frequency and amplitude variations (respectively, termed as *jitter* and *shimmer*) and have substantial inharmonic components due to noncomplete closure of the vocal folds after the so-called *glottal closure instant*; the presence of these noise-like components is correlated with specific events within the pitch period. As a result, the intuitive concept of 'adding a bit of noise' to intonation curves, to amplitude curves or to voiced speech waveforms to 'make them sound more natural' merely leads to more noisy synthetic speech: inharmonicity in voiced speech is not pure randomness. Concatenative synthesis techniques try to deliver the expected intelligibility and naturalness of synthetic speech by gluing together speech chunks which embody natural coarticulation, jitter, shimmer and inharmonicity.

The most famous application of this concept is that of diphone-based speech synthesis. A diphone is a speech segment which starts in the middle of the stable part (if any) of a phoneme and ends in the middle of the stable part of the next phoneme. Diphones therefore have the same average duration as phonemes (about 100 ms), but if a language has N phonemes, it typically has about N^2 diphones.[3] This leads to a typical diphone database size of 1500 diphones (about 3 min of speech, i.e., about 5 Mb for speech sampled at 16 khz/16 bits).

---

3. A bit less in practice because not all diphones are encountered in natural languages.

To create such a synthesiser, one needs to set up a list of required diphones; a corresponding list of words is carefully completed, in such a way that each segment appears at least once. Unfavourable positions, like inside strongly stressed syllables or in strongly reduced (over-coarticulated) contexts, are usually excluded (and will therefore not be perfectly produced at synthesis time). A corpus is then read by a professional speaker and digitally recorded and stored. The elected diphones are then segmented and collected into a diphone inventory (or database). This operation is performed only once.

At runtime, once the synthesiser (Figure 3.1) receives some phonetic input (phonemes, phoneme duration, pitch) from the NLP module, it sets up a list of required diphones, together with their required duration and fundamental frequency contour. The available diphones would only match these prosodic requests by chance because they have generally been extracted from words and sentences which may be completely different from the target synthetic

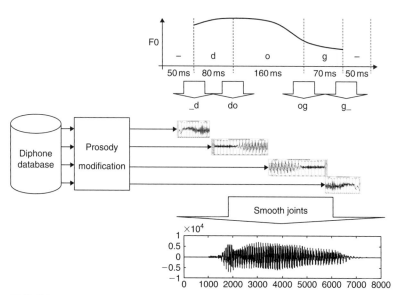

FIGURE 3.1   A diphone-based speech synthesiser. To produce the word 'dog' with given phoneme durations and fundamental frequency, the diphone inventory (or database) is queried for the corresponding sequence of diphones. The duration and pitch of each diphone is modified, and some smoothing is applied on joints.

sentence. Some *prosody modification* is therefore necessary. Additionally, because the diphones to be concatenated have generally been extracted from different words – that is, in different phonetic contexts – the end of one diphone and the beginning of the next one often do not fully match in terms of amplitude and/or spectral envelope. Part of this problem can be solved by *smoothing* individual pairs of successive segments.

Several approaches have been proposed for prosody modification and smoothing. The most common is certainly provided by the time-domain pitch-synchronous overlap-add algorithm [42], which performs these operations in the time domain with very small CPU load. Other methods are based on some form of parametric modelling of speech as in the hybrid harmonic/noise model (HNM) [43]. This model assumes that speech is composed of two additive components, a harmonic component and a noise component, corresponding to a separation of the speech spectrum into two frequency bands. This makes it possible to account for the nonperiodic components of speech, which include frication noise and period-to-period variations of the glottal excitation. Another approach is to edit diphones (once for all) with some HNM-like technique so as to make them more amenable to time-domain overlap add (OLA). This is done in the multiband resynthesis OLA (MBROLA) technique [44].

The acoustic units chosen as building bricks for the concatenative synthesiser should obviously exhibit some basic properties: they should account for as many coarticulatory effects as possible; they should possibly be available with all sorts of durations, pitch contours and voice qualities (so as to alleviate the task of the prosody modification block); they should be easily connectable. Although diphones are a good compromise, partial use of triphones (which embody a complete phoneme) or quadriphones can help increase the final quality [45].

Such units, however, are based on some preestablished phonetic knowledge of coarticulatory or assimilation effects and where they mostly occur. In contrast, nothing truly impedes them from being directly automatically derived on the basis of some automatic analysis of the distribution of their spectral envelopes. This has led to the proposal of a corpus-based, automatic unit set design algorithm termed as *context-oriented clustering* (COC) [46]. COC builds a decision tree for each phoneme, which automatically 'explains' the variability in the acoustic realisation of this phoneme in terms of contextual factors. Starting from a speech database with phonetic labels, it constitutes initial clusters composed of all the speech units (allophones in this case) with the same phonetic label (Figure 3.2). The context tree is then automatically derived by a greedy algorithm. At each step, the cluster with the highest variance (and with more than a minimum number of segments) is split into two classes on the basis of a partition of a *context factor*, and provided the resulting split leads to an average subcluster variance that is significantly lower than the initial one. Possible context factors used (i.e., the factors that are submitted to the tree building algorithm to partition clusters) are the phonetic labels of neighbouring phones up to a given distance from the central one, as well as stress and syntactic boundaries. Broad phonetic classes can also be used as contextual factors, so as to reduce the size of the decision tree and to cope with the problem of data sparsity. It is important to understand that the output of this COC method, when used for concatenative synthesis using fixed inventories, is basically the resulting tree. Once the tree has been built, the content of its leaves (the speech units taken from the speech database and associated to leaves) is reduced to a single representative speech unit (even possibly rerecorded on purpose). This technique can thus be seen as an

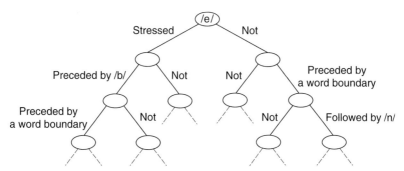

FIGURE 3.2   An example of the context-oriented cluster splitting process for occurrences of the phoneme /e/ (adapted from [46]). As a result of this process, allophones of /e/ are detected and explained.

automatic method for building a list of variants of each phoneme due to specific phonetic/syntactic/stress context.

Although speech synthesis based on the concatenation of units taken from a fixed inventory is clearly intelligible, it cannot really be termed as natural, whichever set of units is chosen. In the next section, we examine how naturalness can be approached.

3.4.3   Unit Selection-Based Synthesis

The lack of naturalness of speech synthesised from a fixed inventory (i.e., with the constraint of having only one instance of each possible concatenative unit stored in the unit database) has two main causes. First, this strategy unavoidably biases the unit recording step towards the choice of a somewhat over-articulated instance, which will fit in most contexts and lead to the best intelligibility score. Additionally, as we have seen in the previous section, signal processing tricks are needed to adapt the pitch and duration of each unit (if only to be able to synthesise stressed and unstressed versions of the same syllable). Some signal degradation is the price to pay for this processing.

It will thus come as no surprise that speech synthesis was about to make a huge step forward when researchers started keeping several instances of each unit and provided means of *selecting*, at runtime, which instance to use for synthesising a given sentence. The foundations of this *unit selection*-based speech synthesis approach was laid in [47] and quickly led to the development of the AT&T NextGen TTS system [48]. At runtime, given a phoneme stream and target prosody for an utterance, the unit selection algorithm selects, from a speech corpus, an optimum set of acoustic units,[4] i.e., the one that 'best' matches the target specifications (Figure 3.3 to be compared to Figure 3.1). For every *target unit* t_i required (for example, for every diphone to be synthesised), the selection algorithm first proposes a list of *candidate units* from the speech database, each in a different context (and in general not exactly in the same context as the target unit).

---

4. Units may still be diphones, although phonemes and subphonetic units such as half-phonemes can be used as alternatives when complete diphones are not available with the required pitch and duration.

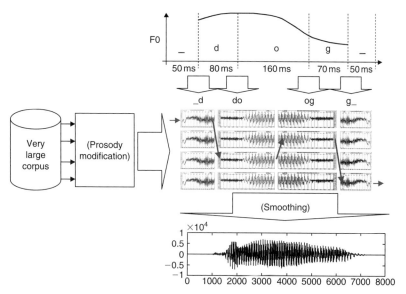

FIGURE 3.3   A schematic view of a unit selection-based speech synthesiser. The prosody modification and smoothing modules have been mentioned between parentheses, because they are not always implemented. As a matter of fact, because this approach uses very large speech corpora, it is often possible to find speech units that naturally join smoothly while exhibiting prosodic features close to what is expected.

The final choice among candidates is based on minimising the sum of two cost functions: a *target cost* $C^t(t_i, u_i)$, which is an estimate of the difference between a candidate unit u_i and the target unit t_i it is supposed to represent, and a *concatenation cost* $C^c(u_{i-1}, u_i)$, which is an estimate of the quality of the joint between candidate units u_{i-1} and u_i for two consecutive targets (Figure 3.4). The best sequence of n units for a given sequence of n targets is the one which minimises to total cost:

$$C(t_1^n, u_1^n) = \sum_{i=1}^{n} C^t(t_i, u_i) + \sum_{i=2}^{n} C^c(u_{i-1}, u_i) + C^c(S, u_1) + C^c(u_n, S),$$

(3.8)

where S denotes the target for a silence. This best sequence is found using a Viterbi search.

When good candidate units cannot be found with the correct pitch and/or duration, prosody modification can be applied. Also, as candidate units usually do not concatenate smoothly (unless a sequence of such candidate units can be found in the speech corpus, perfectly matching the target requirement), some smoothing can be applied. The latest synthesisers, however, tend to avoid prosodic modifications and smoothing, assuming the speech unit inventory is rich enough for the requested units to be available with approximately correct pitch and duration [49]. This assumption is sometimes referred to as *take the best to modify the least*.

The challenges in unit selection synthesis are not only related to the development of efficient target costs [50] and concatenation costs [51] for finding the 'best' path in the candidate unit lattice but also to the definition of optimal speech corpus. As any other expression of a natural language, indeed, speech obeys some form of Zipf's law: *in a corpus of natural language utterances, the frequency of any word is roughly inversely proportional to its rank in the frequency table*. The resulting power law probability distribution implies that speech is composed of a 'large number of rare events' [52]. It was estimated, for instance, that for a diphone database to cover a randomly selected sentence in English with a probability of .75 (meaning that the probability is 0.75 that all the diphones required to produce the sentence are available in the database), 150,000 units are required. This corresponds to a speech corpus of about 5 h. Yet the estimation was based on a limited number of contextual features [53]. Most industrial systems today use from 1 to 10 h of speech, i.e., 150 to 1500 Mb, and reach very high quality synthesis *most of the time*.

It should be noted that the development of larger and larger segmented speech corpora costs a lot. Hence, it is becoming increasingly difficult for companies to deliver tailored voices on demand. Similarly, the voice quality offered by today's TTS systems is somehow uniform. Given the increasing quest for *expressive* speech synthesis, unit selection typically leads to parallel recordings of several speech corpora, each with an identifiable expression; this adds to the cost of a system.

Computing target costs from each target in a large speech database to each candidate unit for that target is found to be time consuming.

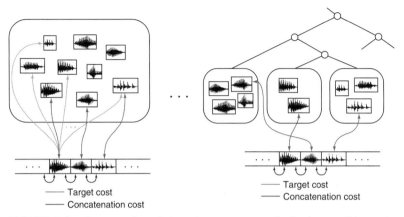

FIGURE 3.4 Context-oriented clustering as a means of selecting candidate units for a given target (after [57]).

By recycling the above-mentioned COC principle and extending it to contextual and prosodic features (pitch, duration and amplitude), it has been possible to drastically reduce the computational load for unit selection [54]. The main novelty here is that once a COC tree is built for a given phoneme not only the tree is retained but also its leaves, i.e., the units that have been clustered based on their feature vectors. At runtime, for each target in the sentence to synthesise, the related COC is browsed so that only the units in the appropriate leaves are considered as candidate units (Figure 3.4). The same idea has been used in the IBM and Microsoft TTS systems in which COC is applied to units associated to HMM states [55, 56].

3.4.4   Statistical Parametric Synthesis

One of the major problems encountered in unit-selection is data sparsity. Synthetic utterances can be perfectly natural if the target sequence of units happens to be available in a continuous sentence in the database, but they can also embody disturbing discontinuities (which can hamper intelligibility) when the required targets are not available. In a word, unit selection lacks a capacity for generalising to unseen data.

In contrast, ASR techniques based on HMM/GMM models have developed such a generalisation property, by making use of context-oriented distribution clustering for parameter tying. This has motivated several research teams to make a very promising step, by considering the speech synthesis problem as a statistical analysis/sampling problem.

In the so-called *HMM-based speech synthesis* approach, implemented in [57], speech is described as the acoustic realisation of a sequence of phonemes. Each phoneme is modelled by a 3-state HMM. Each state emits spectral feature vectors (typically MFCCs) with emission probabilities given by multivariate Gaussians associated to the leaves of a context clustering tree (Figure 3.5). This model is very much inspired by similar work in ASR. Additionally, the pitch and duration of phonemes are modelled by separate clustering trees and Gaussians.

The new idea in HMM-based synthesis is to use this model for generating a stream of parameters (related to a given parametric model of speech), which can then be converted into speech (Figure 3.6).

Let us assume that the models of Figure 3.5 have previously been trained so as to maximise the probability of the speech corpus given the model. The training part is similar to that of an ASR system, except the contextual factors that are used when building the context

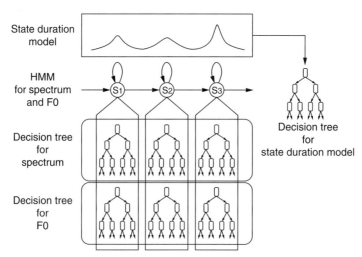

FIGURE 3.5   A statistical parametric model for the spectrum, F0 and duration of a phoneme (from [57]).

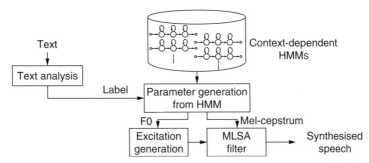

FIGURE 3.6 Speech synthesis using a statistical parametric model (adapted from [58]).

clustering trees incorporate phonetic, stress and syntactic information (which is produced by the TTS system, while it is not accessible to an ASR system). From the target sequence of phonemes (augmented with stress marks and syntactic information), a sentence HMM is constructed (again, as in Figure 3.5) by concatenating CD phoneme HMMs. The state durations of the phoneme HMMs are determined so as to maximise the output probability of state durations given their phonetic and syntactic context. A sequence of spectral parameters $\mathbf{o} = \left[o_1^T, \ldots, o_t^T, \ldots, o_T^T \right]^T$ is then determined in such a way that the output probability of this sequence given the HMM model is maximised, where T is the number of frames in the observation vector sequence (known from the previously estimated state durations).

More precisely, since the duration of each HMM state is known, the sequence of spectral vectors $\mathbf{o}_t^T$ to be produced is associated to a given sentence HMM state sequence $\mathbf{q} = \{q_1, \ldots, q_t, \ldots, q_T\}$. For each such HMM state, the contextual information of the related phoneme is used to find which leaf of the COC tree (associated with this phoneme) will model the production of the state-dependent spectral parameters. The spectral features of the speech segments in each leaf are described by a single multivariate Gaussian distribution $\mathcal{N}\left(\mu_{q_t}, \Sigma_{q_t}\right)$. Then, $P(\mathbf{o}|\mathbf{q}, \lambda)$ is given by

$$P(\mathbf{o}|\mathbf{q}, \lambda) = \prod_{t=1}^{T} \mathcal{N}\left(\mathbf{o}_t | \mu_{q_t}, \Sigma_{q_t}\right). \tag{3.9}$$

Stated as such, however, the sequence of spectral parameters produced by each state of the model will obviously be composed of a sequence of identical spectral parameters: the mean of the underlying multivariate Gaussian (which by nature maximise the Gaussian distribution). The main feature of HMM-based synthesis is thus the use of *dynamic* features, by inclusion of dynamic coefficients in the feature vector. This too is typical of HMM-based ASR techniques but poses a specific problem in HMM-based speech synthesis: how to find a sequence of (static *and* dynamic) spectral features which at the same time has maximum likelihood given the HMM model and the dynamic constraints. This problem has been proved to have an analytical solution [58].

HMM-based speech synthesis currently produces speech with remarkable fluidity (smoothness) but rather poor voice quality. The resulting buzziness is still clearly due to the use of a source/filter model based on linear prediction, as well as to the over-smoothed formant tracks produced by the HMM-based model. It has, however, several important potential advantages over unit-selection. First, its use of context clustering is far more flexible than that of unit-selection because it allows for the creation of separate trees for spectral parameters, F0 and duration. Second, its coverage of the acoustic space is better, given the generative capability of the HMM/COC/Gaussian models. Even more importantly, it embodies a complete model of natural speech with very limited footprint (1 MB). Last but not least, it provides a natural framework for voice modification and conversion.

3.5 CONCLUSIONS

HMM-based automatic speech and speaker recognition is currently a dominant paradigm. The technology has been maturing to a level that allows commercial and viable deployments. Regarding ASR, word level error rates vary, depending on the complexity of the task, but have been reported to be as low as 5% for large vocabulary dictation in clean conditions, 10% for transcribing spontaneous speech from broadcast news and 20% for transcribing spontaneous telephone conversations. Regarding speaker verification, EERs as low as 2% have been reported. Technology is still far away from human capabilities however. State-of-the art systems are very sensitive to both intrinsic (for instance, non-native speech) and extrinsic (background noise) sources of variation. For instance, word error rates are still as high as 10% for digit strings recognition in a car driven on a highway and using a microphone positioned in the dashboard, while it is below 1% when the noise level is very low. Current research is nevertheless attempting to tackle these vulnerabilities, as well as the assumptions and approximations inherent in current technology, with a long-term goal of reaching human-like capabilities. This will hopefully be evidenced by an increase in the range of applications that will be supported by these technologies in the near future.

For speech synthesis, today's winning paradigm in the industry[5] is still unit-selection, while more rudimentary diphone-based systems are still used for low cost consumer products. Statistical parametric synthesis offers very promising results and allows a degree of flexibility in speech control that is not possible with unit selection. It should quickly supersede diphone-based synthesis in industrial applications for which a small footprint is a requirement.

REFERENCES

1. V. Tyagi, C. Wellekens, H. Bourlard, On variable-scale piecewise stationary spectral analysis of speech signals for ASR. Proceedings of Interspeech 2005, 2005, pp. 209–212.
2. K.K. Paliwal, Usefulness of phase in speech processing Proceedings of IPSJ Spoken Language Processing Workshop, 2003, pp. 1–6.
3. H. Bourlard, N. Morgan, Connectionist Speech Recognition: A Hybrid Approach. Kluwer Academic Publishers, 1994.
4. G. Zweig, Speech Recognition with Dynamic Bayesian Networks, Ph.D. Thesis. U.C., Berkeley, CA, 1988.
5. C. Leggetter, P. Woodland, Maximum likelihood linear regression for speaker adaptation of continuous density Hidden Markov Models, Comput. Speech Lang. 9 (2) (1995) 171–185.

---

5. Some speech synthesisers are also available for free for research purposes. See for instance the MBROLA (http://tcts.fpms.ac.be/synthesis/mbrola.html), FESTIVAL (http://www.cstr.ed.ac.uk/projects/festival/), FestVox (http://festvox.org/) or FreeTTS (http://freetts.sourceforge.net/docs/) projects. See also TTSBOX, the tutorial MatlabTM toolbox for TTS synthesis (http://tcts.fpms.ac.be/projects/ttsbox/).

6. M.J.F. Gales, Cluster adaptive training of hidden Markov models, IEEE Trans. Speech Audio Process. 8 (4) (2000) 417–428.

7. H. Yamamoto, S. Isogai, Y. Sagisaka, Multi-class composite n-gram language model, Speech Commun. 41 (2–3) (2003) 369–379.

8. X. Huang, A. Acero, H.-W. Ho, Spoken Language Processing: A Guide to Theory, Algorithm, and System Development. Prentice-Hall, 2001.

9. P. Heracleous, Y. Nakajima, A. Lee, H. Saruwatari, K. Shikano, Accurate hidden markov models for non-audible murmur (nam) recognition based on iterative supervised adaptation, in: Proceedings of ASRU 2003, U.S. Virgin Islands, 2003, pp. 73–76.

10. O.M. Strand, T. Holter, A. Egeberg, S. Stensby, On the feasibility of ASR in extreme noise using the parat earplug communication terminal, 2003.

11. M. Graciarena, H. Franco, K. Sonmez, H. Bratt, Combining standard and throat microphones for robust speech recognition, IEEE Signal Process. Lett. 10 (3) (2003) 72–74.

12. Y. Zheng, Z. Liu, Z. Zhang, M. Sinclair, J. Droppo, L. Deng, et al., Air- and bone-conductive integrated microphones for robust speech detection and enhancement. Proceedings of ASRU 2003, U.S. Virgin Islands, 2003, pp. 249–254.

13. S. Dupont, C. Ris, Combined use of close-talk and throat microphones for improved speech recognition under non-stationary background noise. Proceedings of Robust 2004 (Workshop (ITRW) on Robustness Issues in Conversational Interaction), Norwich, UK, August 2004.

14. G. Potamianos, C. Neti, J. Luettin, I. Matthews, Audio-visual automatic speech recognition: an overview. Issues in Visual and Audio-Visual Speech Processing, 2004.

15. S.J. Cox, S. Dasmahapatra, High level approaches to confidence estimation in speech recognition, IEEE Trans. Speech Audio Process, 10 (7) (2002) 460–471.

16. H. Jiang, Confidence measures for speech recognition: a survey, Speech Commun. 45 (4) (2005) 455–470.

17. P.C. Loizou, I. Cohen, S. Gannot, K. Paliwal, Special issue on speech enhancement. Speech Commun. Spec. Issue. 49 (7–8) (2007) 527–529.

18. F. Nolan, The Phonetic Bases of Speaker Recognition. Cambridge University Press, 1983.

19. P.L. Garvin, P. Ladefoged, Speaker identification and message identification in speech recognition. Phonetica, 9 (1963) 193–199.

20. M. Benzeghiba, R.D. Mori, O. Deroo, S. Dupont, T. Erbes, D. Jouvet, et al., Automatic speech recognition and speech variability: a review. Speech Commun. 49 (10–11) (2007) 763–786.

21. J.R. Glass, A probabilistic framework for segment-based speech recognition, Comput. Speech Lang. 17 (2–3) (2003) 137–152.

22. T. Hain, Implicit modelling of pronunciation variation in automatic speech recognition, Speech Commun. 46 (2) (2005) 171–188.

23. P.B. de Mareuil, M. Adda-Decker, A. Gilles, L. Lori, Investigating syllabic structures and their variation in spontaneous French. Speech Commun. 46 (2) (2005) 119–139.

24. E. Fosler-Lussier, W. Byrne, D. Jurafsky, Pronunciation modelling and lexicon adaptation, Speech Commun. 46 (2) (2005).

25. K.-T. Lee, Dynamic pronunciation modelling using phonetic features and symbolic speaker adaptation for automatic speech recognition, PhD Thesis, Swiss Federal Institute of Technology, 2003.

26. L. Lee, R.C. Rose, Speaker normalisation using efficient frequency warping procedures, Proceedings of ICASSP, 1996.

27. A. Mertins, J. Rademacher, Vocal tract length invariant features for automatic speech recognition, Proceedings of ASRU, 2005, pp. 308–312.

28. D. Elenius, M. Blomberg, Comparing speech recognition for adults and children. Proceedings of FONETIK 2004 (2004) 156–159.

29. J. Zheng, H. Franco, A. Stolcke, Rate of speech modelling for large vocabulary conversational speech recognition. Proceedings of ISCA tutorial and research workshop on automatic speech recognition: challenges for the new Millenium, 2000, pp. 145–149.

30. D. Ververidis, C. Kotropoulos, Emotional speech recognition: resources, features, and methods, Speech Commun. 48 (9) (2006) 1162–1181.

31. G. Zweig, Speech Recognition with Dynamic Bayesian Networks. PhD thesis, University of California, Berkeley, 1998.

32. S. Dupont, C. Ris, Multiple acoustic and variability estimation models for ASR. Proceedings of SRIV 2006 International Workshop on Speech Recognition and Intrinsic Variation, Toulouse, France, May 2006.

33a. F. Bimbot, J. Bonastre, C. Fredouille, G. Gravier, I, Magrin-Chagnolleau, S. Meignier, et al., A tutorial on text-independent speaker verification. EURASIP J. Appl. Signal Process. (2004) 430–451.

33b. L. Marya, B. Yegnanarayana, Extraction and representation of prosodic features for language and speaker recognition, Speech Commun. 50 (10) (2008) 782–796.

34. F. Bimbot, A tutorial on text-independent speaker verification. EURASIP J. Appl. Signal Process. 451 (2004) 430–451.

35. D.A. Reynolds, Speaker verification using adapted Gaussian mixture models, Digit. Signal Process. 10 (2000) 19–41.

36. O. Thyes, R. Kuhn, P. Nguyen, J.-C. Junqua, Speaker identification and verification using eigenvoices. Proceedings ICSLP-2000, 2 (2000) 242–245.

37. A. Stolcke, L. Ferrer, S. Kajarekar, E. Shriberg, A. Venkataraman, MLLR transforms as features in speaker recognition. Proceedings of the 9th European Conference on Speech Communication and Technology (Interspeech 2005-Eurospeech 2005), Lisboa, Portugal, 2005, pp. 2425–2428.

38. V. Wan, W.M. Campbell, Support vector machines for verification and identification. Neural Networks for Signal Processing X, Proceedings of the 2000 IEEE Signal Processing Workshop, vol. 2000, 2000, pp. 775–784.

39. K. Kumar, Q. Wu, Y. Wang, M. Savvides, Noise robust speaker identification using Bhattacharya distance in adapted Gaussian models space. EUSIPCO-2008, Lausanne, Switzerland, 2008.

40. P. Taylor, Text-to-Speech Synthesis. Cambridge University Press, Cambridge 2009.

41. K. Silverman, M. Beckman, J. Pitrelli, M. Ostendorf, C. Whightman, P. Price, et al., ToBI: a standard for labeling English prosody, Proceedings of the International Conference on Spoken Language Processing (ICSLP'92), 1992, pp. 867–870.

42. E. Moulines, F. Charpentier, Pitch Synchronous waveform processing techniques for Text-To-Speech synthesis using diphones. Speech Commun. 9 (1990) 5–6.

43. Y. Stylianou, Applying the Harmonic plus Noise Model in Concatenative Synthesis, IEEE Trans. Speech Audio Process. 9 (1) (2001) 21–29.

44. T. Dutoit, H. Leich, MBR-PSOLA: text-to-speech synthesis based on an MBE resynthesis of the segments database, Speech Commun. 13 (13) (1993) 435–440.

45. D. Bigorne, O. Boeffard, B. Cherbonnel, F. Emerard, D. Larreur, J.L.L. Saint-Milon, et al., Multilingual PSOLA text-to-speech system, Proc. Int. Conf. Acoust. Speech Signal Process. 2 (1993) 187–190.

46. S. Nakajima, Automatic synthesis unit generation for English speech synthesis based on multi-layered context oriented clustering, Speech Commun. (14) (1994) 313–324.

47. A.J. Hunt, A.W. Black, Unit selection in a concatenative speech synthesis system using a large speech database. Proceedings of the International Conference on Acoustics, Speech, and Signal Processing (ICASSP'96), vol. 1, 373–376. Atlanta, Georgia, 1996.

48. M. Beutnagel, A. Conkie, J. Schroeter, Y. Stylianou, A. Syrdal, The AT&T next-gen TTS system, in: Proceedings of the Joint Meeting of ASA, EAA, and DAGA, Berlin, Germany, 1999, pp. 18–24.

49. N. Campbell, A. Black, Prosody and the selection of source units for concatenative synthesis, in: J. van Santen, R. Sproat, J. Olive, J. Hirshberg, (Eds.), Progress in Speech Synthesis. Springer Verlag, 1995.
50. P. Taylor, A.W. Black, Speech synthesis by phonological structure matching. Proceedings of Eurospeech'99, 99 (1999) 623–626.
51. J. Vepa, S. King, Join cost for unit selection speech synthesis, in: A. Alwan, S. Narayanan, (Eds.), Speech Synthesis, 2004.
52. B. Möbius, Rare events and closed domains: two delicate concepts in speech synthesis. 4th ISCA Workshop on Speech Synthesis, 2001 (2001) pp. 41–46.
53. J.P.H. van Santen, Combinatorial issues in text-to-speech synthesis. Proceedings of the European Conference on Speech Communication and Technology (Rhodos, Greece), 5 (1997) 2511–2514.
54. A.W. Black, P. Taylor, Automatically clustering similar units for unit selection in speech synthesis. Proceedings of Eurospeech, 1997, p. 2.
55. R.E. Donovan, E.M. Eide, The IBM trainable speech synthesis system. Proceedings of the International Conference on Spoken Language Processing (Sydney, Australia), vol. 5 (1998) pp. 1703–1706.
56. X. Huang, A. Acero, J. Adcock, H. Hon, J. Goldsmith, J. Liu, et al., Whistler: a trainable text-to-speech system. Proceedings ICSLP-96, Philadelphia, 1996, pp. 659–662.
57. K. Tokuda, H. Zen, A. Black, An HMM-based speech synthesis system applied to English, Proceedings IEEE Speech Synthesis Workshop, 2002.
58. H. Zen, K. Tokuda, T. Kitamura, An introduction of trajectory model into HMM-based speech synthesis, Proceedings Speech Synthesis Workshop SSW5, Pittsburgh, 2005.

Chapter 4

Natural Language and Dialogue Processing

Olivier Pietquin
Supélec Campus de Metz, France

4.1 INTRODUCTION

Natural language processing (NLP) traditionally involves the manipulation of written text and abstract meanings, aiming at relating one to the other. It is of major interest in multimodal interfaces since

Multimodal Signal Processing, ISBN: 9780123748256

written inputs and outputs are important means of communication but also because speech-based interaction is often made via textual transcriptions of spoken inputs and outputs. Whatever the information supports, the interaction is based on the assumption that both participants do understand each other and produce meaningful responses. Speech recognition and speech synthesis are, therefore, mandatory but not sufficient to build a complete voice-enabled interface. Humans and machines have to manipulate meanings (either for understanding or producing speech and text) and to plan the interaction. In this contribution, NLP will be considered under three points of view. First, natural and spoken language understanding (NLU or SLU) aiming at extracting meanings from text or spoken inputs. Second, natural language generation (NLG) that involves the transcription of a sequence of meanings into a written text. Finally, dialogue processing (DP) that analyses and models the interaction from the system's side and feeds NLG systems according to extracted meanings from users' inputs by NLU.

4.2 NATURAL LANGUAGE UNDERSTANDING

It is the job of an NLU system to extract meanings of text inputs. In the case of SLU, previous processing systems (such as automatic speech recognition: ASR, see Chapter 3) are error-prone and can add semantic noise such as hesitations, stop words, etc. This has to be taken into account. To do so, some NLU systems are closely coupled to the ASR system, using some of its internal results (Nbest lists, lattices or confidence scores). Other NLU systems maintain several hypotheses so as to propagate uncertainty until it can be disambiguated by the context.

Assuming that the input is a correct word sequence, most of NLU systems can be decomposed in three steps: syntactic parsing, semantic parsing and contextual interpretation. In the following, a brief overview of each step is given. For further details about the basic ideas and methods of NLU, readers are invited to refer to [1].

4.2.1 Syntactic Parsing

Before trying to extract any meaning out of a sentence, the syntactic structure of this sentence is generally analysed: the function of each word (part of speech), the way words are related to each other, how they are grouped into phrases and how they can modify each other. It helps resolving some ambiguities as homographs (homophones) having different possible functions. For instance, the word 'fly' can be a noun (the insect) or a verb and the word 'flies' can stand for the plural form of the noun or an inflexion of the verb as shown on Figure 4.1.

Most syntactic representations of language are based on the notion of context-free grammars (CFG) [2]. Sentences are then split in a hierarchical structure (Figure 4.1). Most early syntactic parsing algorithms, aiming at creating this parse tree, were developed with the goal of analysing programming languages rather than natural language [3]. Two main techniques for describing grammars and implementing parsers are mainly used: *context-free rewrite rules* and *transition networks* [4].

For instance, a grammar capturing the syntactic structure of the first sentence in Figure 4.1 can be expressed by a set of rewrite rules as follows:

1. $S \rightarrow NP \ VP$
2. $NP \rightarrow A \ N$
3. $VP \rightarrow V$
4. $A \rightarrow The$
5. $N \rightarrow fly$
6. $V \rightarrow flies$

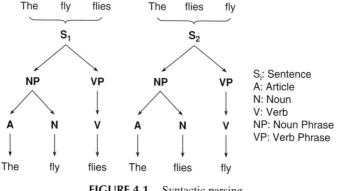

FIGURE 4.1   Syntactic parsing.

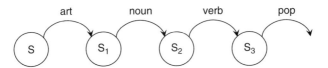

FIGURE 4.2   Transition network grammar.

The BNF (Backus–Nauer Form) [5] notations are often used to express those rules. First three rules are called *non-terminal rules* because they can still be decomposed and others are called *terminal rules* (or terminal symbols). Rewrite rules are very similar to those used in logical programming (like PROLOG). This is why logical programming has been widely used for implementing syntactic parsers [6].

The aforementioned grammar can also be put into the form of the state-transition network (STN) shown in Fig. 4.2.

Natural languages often involve restrictions in the combinations of words and phrases. Many forms of agreements exist such as number agreement, subject–verb agreement, gender agreement, etc. For instance, the noun phrase 'a cats' is not correct in English, as it does not satisfy the number agreement restriction. The previously described formalisms can be extended to handle these restrictions. Constituents are then associated to features (like number, gender, etc.) and this leads to *augmented grammars*. There are augmented CFGs and augmented transition networks (ATN).

CFGs and ATNs have been the topic of extensive research. Yet, they are still limited because often handcrafted and deterministic. Coping with ambiguities is made uneasy when several interpretations are possible. With the emergence of data-driven and statistical techniques came other solutions to the parsing problem [7]. For instance, the part-of-speech tagging problem can be expressed as the problem of selecting the most likely sequence of syntactic categories $(C_1, \ldots, C_T)$ for the words $(w_1, \ldots, w_T)$ in a sentence, that is the sequences $(C_1, \ldots, C_T)$ that maximises the probability:

$$P(C_1, C_2, \ldots, C_T | w_1, w_2, \ldots, w_T) \tag{4.1}$$

Using Bayes' rule:

$$P(C_1, C_2, \ldots, C_T | w_1, w_2, \ldots, w_T)$$
$$= \frac{P(w_1, w_2, \ldots, w_T | C_1, C_2, \ldots, C_T) P(C_1, C_2, \ldots, C_T)}{P(w_1, w_2, \ldots, w_t)} \quad (4.2)$$

The solution to the syntactic parsing problem is:

$$\underset{C_1, C_2, \ldots, C_T}{\mathrm{argmax}} \; P(w_1, w_2, \ldots, w_T | C_1, C_2, \ldots, C_T) P(C_1, C_2, \ldots, C_T),$$

$$(4.3)$$

since the the denominator remains constant. This is generally not computable because probabilities involved in this expression can hardly be analytically expressed and even difficult to estimate from data in the case of long sentences.

Two assumptions are then often made. First, let's assume that $P(C_1, \ldots, C_T)$ can be approximated by an n-gram model (see Chapter 3, Section 3.2.3). That is assuming that the probability of occurrence of category C_i only depends on the $n-1$ previous categories:

$$P(C_i | C_{i-1}, \ldots, C_0) = P(C_i | C_{i-1}, \ldots, C_{i-(n-1)}) \quad (4.4)$$

A bigram model is often adopted and only $P(C_i | C_{i-1})$ is estimated, which leads to:

$$P(C_1, C_2, \ldots, C_T) = P(C_1) \prod_{i=2}^{T} P(C_i | C_{i-1}) \quad (4.5)$$

Second, let's assume that a word appears in a category independently of the word in the preceding and the succeeding categories:

$$P(w_1, w_2, \ldots, w_T | C_1, C_2, \ldots, C_T) = \prod_{i=1}^{T} P(w_i | C_i) \quad (4.6)$$

In this case, each $P(w_i | C_i)$ can be compared to the emission probability of the ASR problem, while each $P(C_i | C_{i-1})$ can be compared to a transition probability and Hidden Markov Models (HMM) can therefore serve as a probabilistic model. The part-of-speech tagging problem can be translated into the search of the sequence $(C_1, \ldots, C_T)$ that maximises:

$$P(w_1|C_1)P(C_1) \prod_{i=2}^{T} P(w_i|C_i)P(C_i|C_{i-1}) \qquad (4.7)$$

All the probabilities in the aforementioned equation can be estimated using a manually annotated data corpus (even if unaligned, thanks to the EM algorithm) and the problem can then be solved with the same tools used for solving the ASR problem (a Viterbi algorithm, for instance).

When dealing with spoken language, syntactic parsing is often inefficient because of speech recognition errors, hesitations, etc. [8]. This is why semantic-only-based methods have been investigated for SLU.

4.2.2 Semantic Parsing

The role of a semantic parser is to extract the context-independent meaning of a written sentence. For instance, the noun phrase 'video cassette recorder' has a single meaning and refers to the device that records and plays back video tapes, whatever the context. In the same way, albeit the word 'flies' is ambiguous, once it has been correctly identified as a verb by the syntactic parser it does not need any contextual interpretation to reveal its meaning (yet, syntactic parsing was required).

However, in a large number of cases, a single word can endorse several meanings that cannot be disambiguated by a syntactic parsing. The other way around, a same meaning can also have several realisations (synonyms). Often, some meanings of a specific word can be eliminated at the sentence level because they do not fit the direct context of the word. For instance, in the sentence 'John wears glasses', the word 'glasses' means spectacles and not receptacles containing a liquid, because John cannot 'wear' receptacles. Remaining ambiguities are context-dependent and will be considered in the next sub-section.

Given this, it appears that semantic interpretation resembles a classification process aiming at categorising words or groups of words in classes regrouping synonyms. The set of classes in a particular representation of the world (or at least of a domain) is called its *taxonomy*, while the relationships between those classes is the *ontology*. Such classifications have been of interest for a very long time and arise in Aristotle's writings in which he suggested the following major classes: substance, quantity, quality, relation, place, time, position, state, action and affection. Some information has to be enclosed in the semantic representation. For example, the verb is often described as the word in a sentence that expresses the action, while the subject is described as the actor. Other roles can be defined and are formally called *thematic roles*. Semantic representations should therefore enclose information about thematic roles.

Utterances are not always used to make simple assertions about the world. The sentence 'Can you pass the salt?' is not a question about the ability of someone to pass the salt but rather a request to actually pass the salt. Some utterances have therefore the purpose of giving rise to reactions or even of changing the state of the world. They are the observable performance of communicative actions. Several actions can be associated to utterances like assertion, request, warning, suggestion, informing, confirmation, etc. This is known as the *Speech Act* theory and has been widely studied in the field of philosophy [9]. When occurring in a dialogue, they are often referred to as *Dialog Acts*. The Speech Act theory is an attempt to connect language to goals. In the semantic representation of an utterance, speech acts should therefore be associated (often one speech act per phrase). Computational implementations of this theory have been developed early [10]. Philosophy provided several other contributions to natural language analysis and particularly in semantics (like the lambda calculus, for example) [11].

From this, Allen proposes the logical form [1] framework to map sentences to context independent semantic representations. Although it is not exploited in every system, numerous currently used semantic representation formalisms are related to logical forms in one way or another. Allen also proposes some methods to automatically transfer from a syntactic representation to the corresponding logical form.

Statistical data driven methods such as HMM previously used in speech recognition and syntactic parsing have also been applied to semantic parsing of spoken utterances [12]. Semantic parsing can indeed be expressed as the problem of selecting the most likely sequence of concepts $(C_1, \ldots, C_t)$ for the words $(w_1, \ldots, w_T)$ in a sentence. This problem can be solved considering that the concepts are hidden states of a stochastic process while the words are observations. Notice that this technique bypasses the syntactic parsing and can be adapted to SLU. More recently, using the same idea, other generative statistical models as dynamic Bayesian networks were used to infer semantic representations [13]. The advantage of such models is that they do not need to be trained on manually aligned data that is time-consuming to obtain.

One disadvantage of generative methods is that they are not trained discriminatively like supervised methods. They also make some independent assumptions over the features. This results generally in lower performances in classification tasks. Discriminative methods such as support vector machines (SVM) have therefore been applied to the problem of semantic parsing [14]. But these last methods require accurate labelled data that are generally not available. Consequently, combinations of discriminative and generative models trained on unaligned data has recently been introduced [15, 16].

Other statistical methods such as decision trees and classification and regression trees (CART) are also often part of NLU systems for semantic parsing [17]. They extract a so-called semantic- or parse-tree from a sentence and aim at finding the most probable tree that fits the sentence.

4.2.3 Contextual Interpretation

Contextual interpretation takes advantage of information at the discourse level to refine the semantic interpretation and to remove remaining ambiguities. Here, the term discourse defines any form of multi-sentence or multi-utterance language. Three main ambiguities can be resolved at the discourse level:

Anaphors are parts of a sentence that typically replace a noun phrase (e.g., the use of the words 'this', 'those', …). Several types of anaphors can be cited such as intrasentence (the reference is in the same sentence), intersentence (the reference is in a previous sentence) and surface anaphors (the reference is evoked but not explicitly referred to in any sentence of the discourse).

Pronouns are particular cases of anaphors and typically refer to noun phrases. Several specific studies have addressed this kind of anaphora.

Ellipses involve the use of clauses (parts of discourse that can be considered as stand-alone) that are not syntactically complete sentences and often refer to actions or events. For instance, the second clause (B) in the following discourse segment is an ellipsis:

- A. John went to Paris.
- B. I did too.

Although there exist a wide range of techniques for resolving anaphors, they are almost all based on the concept of *discourse entity* (DE) [18]. A DE can be considered as a possible antecedent for an unresolved ambiguity in the local context and it is typically a possible antecedent for a pronoun. The system maintains a DE list which is a set of constants referring to objects that have been evoked in previous sentences and can subsequently be referred to implicitly. A DE is classically generated for each noun phrase.

Simple algorithms for anaphora resolutions based on DE history exist (the last DE is the most likely to be referred to). Yet, the most popular methods are based on a computational model of discourse focus [19] that evolved into the centering model [20]. These theories rely on the idea that the discourse is organised around an object (the centre) that the discourse is about, and that the centre of a sentence is often pronominalised. The role of the system is then to identify the current centre of the discourse and to track centre moves.

4.3   NATURAL LANGUAGE GENERATION

Natural language generation (NLG) aims at producing understandable text from non-linguistic representations of information (concepts). Research in this field started in the 1970s, that is later than NLU or ASR. Thus, reference books are quite recent [21]. Applications of NLG are automatic documentation of programming language [22], summarisation of e-mails, information about weather forecasts [23] and spoken dialogue systems [24–26]. Many NLG techniques exist and particularly when the generated text is intended to be used for speech synthesis (concept-to-speech synthesis).

In most industrial spoken dialogue systems, the NLG sub-system is often very simple and can be one of the following:

Pre-recorded prompts sometimes a real human voice is still preferred to text-to-speech (TTS) systems because it is more natural. This is only possible if the set of concepts that have to be managed is small enough. In this case, a simple table mapping concepts to corresponding audio records is built. This technique has several drawbacks as the result is static and recording human speech is often expensive.

Human authoring where the idea is approximately identical except that the table contains written text for each possible concept sequence. The text is subsequently synthesised by a TTS system. This solution is more flexible as written text is easier to produce and modify but still needs human expertise. This technique is still extensively used.

Although using one of those techniques is feasible and widely done in practice, it is not suitable for certain types of dialogues like tutoring, problem-solving or troubleshooting for example. In such a case, the system utterances should be produced in a context-sensitive fashion, for instance by pronominalising anaphoric references, and by using more or less sophisticated phrasing or linguistic style [27, 26] depending on the state of the dialogue, the expertise of the user, etc. Therefore, more complex NLG methods are used and the process is commonly split into three phases [21]: document planning, microplanning and surface realisation. This first phase is considered as language independent while the two following are language dependent.

4.3.1 Document Planning

NLG is a process transforming high-level communicative goals into a sequence of communicative acts (Speech Acts), which accomplish the initial communicative goals. The job of the document-planning phase (or text-planning phase) is to break high-level communicative goals into structured representations of atomic communicative goals that can be attained with a single speech act (in language, by uttering a single clause). Document planning results in a first overview of the document structure to be produced. It also creates an inventory of the information contained in the future document. It is very task-dependent and techniques used for document planning are closely related to expert systems techniques.

4.3.2 Microplanning

During the microplanning (or sentence planning) phase, abstract linguistic resources are chosen to achieve the communicative goals (lexicalisation). For instance, a particular verb is selected for expressing an action. An abstract syntactic structure for the document is then built (aggregation). Microplanning is also the phase where the number of generated clauses is decided so as to produce language with improved naturalness, a last process focuses on referring expressions generation (e.g., pronoun, anaphora). Indeed, if several clauses of the generated document refer to the same discourse entity, the entity can be pronominalised after the first reference. The result is a set of phrase prototypes (or *proto-phrases*).

Microplanning can be achieved using several methods. Some of them are based on *reversed parsing*, that is semantic grammars used for parsing in NLU are used in a generative manner to produce document prototypes [28]. Most of probabilistic grammar models are generative models that can be used in this purpose. The idea seems attractive but the development of semantic grammars suitable for both NLG and NLU appeared to be very tricky and in general, same meanings will always lead to the same generated proto-phrases (which is unnatural). Other techniques are based on grammar specifically designed for NLG purposes. In spoken dialogue systems, template-based techniques are widely used in practice [29] but learning techniques such as boosting or even reinforcement learning have been recently successfully applied to sentence planning [30, 26].

4.3.3 Surface Realisation

During surface realisation, the abstract structure (proto-phrases) built during microplanning are transformed into surface linguistic utterances by adding function words (such as auxiliaries and determiners), inflecting words, building number agreements, determining word order, etc. The surface realisation process even more than the previous one is strongly language dependent and uses resource specific to the target language. This phase is not a planning phase in that it only executes decisions made previously, using grammatical information about the target language.

4.4 DIALOGUE PROCESSING

Human–human dialogues are mostly multimodal mixing speech, gesture, drawings but also semantic and pragmatic knowledge. When a person engages a conversation with another, information coming from all senses are combined to background knowledge to understand the other interlocutor. Although we focus here on speech-based dialogues, it is still without loss of generality. We also focus on goal-directed dialogues, that is dialogues in which both participants collaborate to achieve a goal. Social dialogue is not in the scope of this contribution. A Spoken Dialog System (SDS) is not only the combination of speech and NLP systems. The interaction has to be managed in some way. The role of the dialogue management component in a man–machine interface is to organise the interaction in terms of sequencing. A man–machine dialog is here considered as a turn-taking process in which information is transferred from one participant to the other through (possibly noisy) channels (ASR, NLU, NLG and TTS). One participant is the user, the other being the dialogue manager (DM, see Figure 4.3). The DM has thus to model the dialogue progress over time but also its structure and the discourse structure.

4.4.1 Discourse Modelling

Human–human dialogue has been the topic of intensive investigations since the beginning of artificial intelligence research. One aim of these investigations was the development of artificial spoken dialogue systems. Discourse modelling aimed at developing a theory of dialogue, including, at least, a theory of cooperative task-oriented dialogue, in which the participants are communicating in the aim of achieving some goal-directed task. Although there is no common agreement on the fact that human–human dialogues should serve as a model

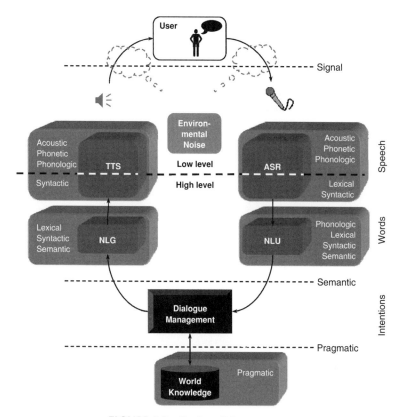

FIGURE 4.3   Spoken dialogue system.

for human–machine dialogues (users' behaviour being often adapted when talking to machines [31, 32]), several approaches to modelling human–machine dialogues coming from studies about human–human dialogues can be cited such as dialogue grammar models, plan-based models, conversational game models and joint action models.

Dialogue grammars are an approach with a long history. This approach considers that regularities exist in a dialogue sequencing and that some Speech Acts (or Dialogue Acts) can only follow some others [33]. For example, a question is generally followed by an answer. In the vein of the NLU process, a set of rules imposes sequential and hierarchical constraints on acceptable dialogues, just as syntactic grammar rules state constraints on grammatically acceptable utterances. The major drawback of this approach (and what makes it so simple) is that it considers that the relevant history for the current dialogue state is only the previous sentence (the previous Speech Act or Dialogue Act). For this reason only few systems were successfully based on this model [32].

Plan-based models are founded on the observation that humans do not just perform actions randomly, but rather they plan their actions to achieve various goals, and in the case of communicative actions (Speech Acts), those goals include changes to the mental state of listeners. Plan-based models of dialogue assume that the speaker's Speech Acts are part of a plan, and the listener's job is to uncover and respond appropriately to the underlying plan, rather than just to the previous utterance [34]. In other words, when inferring goals from a sentence, the system should take the whole dialogue into account and interpret the sentence in the context of the plan. Those models are more powerful than dialogue grammars, but goal inference and decision making in the context of plan-based dialogue are sometimes very complex [35]. Nevertheless, plan-based dialogue models are used in SDSs [36].

The *Conversational Game Theory* (CGT) [37] is an attempt to combine the ideas from both plan-based models and dialogue grammars in a same framework. Dialogues are then considered as exchanges called *games*. Each game is composed with a sequence of *moves* that are valid according to a set of rules (similar to grammars) and the overall game has a goal planned by participants (similar to plan-based models). Participants share knowledge (beliefs and goals) during the dialogue and games can be nested (sub-dialogues are possible) to achieve sub-goals. Moves are most often assimilated to Speech or Dialogue Acts [38]. The CGT defines quite formally the moves allowed for each of the participants at a given state in the game (according to the rules and the goal) and simulation of dialogues is made possible as well [37]. Computational versions of CGT have been successfully applied to dialogue systems [39].

The previous approaches considered a dialogue as a product of the interaction of a plan generator (the user) and a plan recogniser (the dialogue manager) working in harmony, but it does not explain why participants ask clarification questions, why they confirm, etc. This process used by dialogue participants to ensure that they share the background knowledge necessary for the understanding of what will be said later in the dialogue is called *grounding* [40]. According to this, another dialogue model has emerged in which dialogue is regarded as a *joint activity* shared by participants. Engaging in a dialogue requires the interlocutors to have at least a joint agreement to understand each other, and this motivates the clarifications and confirmations so frequent in dialogues. Albeit this family of dialogue models is of a great interest, its complexity makes it difficult to implement [41].

4.4.2   Dialogue Management

Any of the dialogue models described earlier can be used in SDSs to interpret semantics or build internal states, yet the dialogue management can be of several kinds. A dialogue manager implements an interaction strategy aiming at organising the sequencing of systems dialogue acts so as to achieve the common goal of the user and the system. In the following, different dialogue management methods found in the literature are further described. Although not exhaustive, this list is representative of what exists in most current spoken dialogue systems.

Theorem Proving

The *theorem proving* approach to dialogue management has been developed in [42] in the framework of a problem-solving (or troubleshooting) SDS for equipment fixing. The idea underlying this method is that the system tries to demonstrate that the problem is solved (theorem). As in mathematical demonstration, there are several steps to follow. At each step, the system can use *axioms* (things known for sure) or deductions (obtained by logic inference) to move on. If in a given dialogue state, the *World Knowledge* included in the system does not contain an axiom allowing the dialogue manager to go on or if the dialogue manager cannot infer it from other axioms, the axiom is considered as missing (*missing axiom theory*). The system thus prompts the user for more information. If the user is not able to provide information, a new theorem has to be solved: 'the user is able to provide relevant information'. A tutoring sub-dialogue then starts. This technique has been implemented in Prolog as this rule-based method is suitable for logic programming. The major drawback of theorem proving management is that the strategy is fixed as the demonstration steps are written in advance. Also this method is strongly task-dependent and rules can hardly be reused across domains.

Finite-State Machine

In the case of *finite-state* methods, the dialogue is represented like a STN where transitions between dialogue states specify all legal paths through the network. In each state an action is chosen according to the dialogue strategy. Actions can be greeting the user, request information, provide information, ask for confirmation, closing the dialogue, etc. The result of the action leads to a new transition between states. The construction of a dialogue state is a tricky problem. Usually, dialogue states are built according to the information that has been exchanged between the user and the system. In this case, the state is *informational*. For example, if the aim of a dialogue system is to retrieve the first and family names of a person, four dialogue states can be built:

- both names are unknown,
- only the first name is known,
- only the family name is known,
- both names are known.

The main drawback of these methods is that all possible dialogues, that is all paths in the state space, have to be known and described by the designer. The structure is mostly static resulting in inflexible and often system-led behaviours. Nevertheless, state-transition methods have been widely used in dialogue systems but above all in various toolkits and authoring environments [43] for several reasons. It is an easy manner for modelling dialogues that concern well-structured tasks that can be mapped directly on to a dialogue structure. Finite-state methods are easier to understand and more intuitive for the designer as they provide a visual, global and ergonomic representation. Scripting languages such as VoiceXML [44] can be very easily used for representing STNs. As discussed in [43], many applications such as form-filling, database querying or directory interrogation are more successfully processed this way. These applications are only the most popular ones. Eventually, nested sub-dialogues may help to gain in flexibility in this type of method.

Form Filling

Form filling methods, also called *frame-driven*, *frame-based* or *slot-based* methods, are suitable when the goal of the application consists in a one-way transfer of information from the user to the SDS. The information has to be represented as a set of attribute-value pairs. The attribute-value structure can be seen as a form and the user should provide values for each field (attribute, slot, frame) of the form. Each empty field is then eligible for a prompt from the system to the user. The dialogue strategy aims at filling completely the form and to retrieve values for all the fields. Each field is given a priority defining the sequence in which the user is prompted [45]. Each of the user's utterance has then to be processed to provide an attribute-value representation of its meaning.

Self-Organised Management

Unlike aforementioned techniques, the *self-organised* management does not require all the dialogue paths to be specified in advance. Each action of the system and reaction of the user contributes to build a new configuration to which a particular behaviour is associated. There is no need to know how the configuration occurred. This is generally referred to as event-driven dialogue management. One famous attempt to use this kind of dialogue management method was the Philips Speechmania software based on the HDDL language (Harald's Dialogue Description Language, from the first name of its creator) [46]. Other attempts were made to use event-driven dialogue management [47] but the complexity of development of such systems overcame their potentialities.

Markov Decision Processes and Statistical Optimisation

Because rapid development and reusability of previously designed systems are almost impossible with aforementioned management techniques but also because objective assessment of performance is very difficult, machine learning methods for automatic search of optimal interaction policies have been developed during the last decade. This development is also due to the fact that a dialogue strategy has to take into account varying factors that are difficult to model – like performances of the subsystems (like ASR, NLU, NLG, TTS, etc.), the nature of the task (e.g., form filling [48], tutoring [49], robot control, or database querying [50]) and the user's behaviour (e.g., cooperativeness, expertise [51]).

The main trend of research concerns reinforcement learning [52] methods in which a spoken dialogue is modelled as a *Markov decision process* (MDP) [51, 53–55] or a *partially observable MDP* [56, 57]. One concern for such approaches is the development of appropriate dialogue corpora for training and testing. However, the small amount of data generally available for learning and testing dialogue strategies does not contain enough information to explore the whole space of dialogue states (and of strategies). Therefore, dialogue simulation is most often required to expand the existing dataset and man–machine spoken dialogue stochastic modelling and simulation has become a research field in its own right. Particularly, user simulation is an important trend of research [48, 54, 58–61] but also NLU [62] and ASR [63, 64] simulation.

4.4.3 Degrees of Initiative

A major issue of dialogue strategy design is the degree of initiative let to the user in each dialogue state. Roughly speaking, the dialogue management will be easier if the control of the dialogue flow is completely let to the system, while a user will probably be more satisfied if he/she can be over-informative and take initiatives. Three degrees of initiative can be distinguished as follows:

System-led   the system controls completely the dialogue flow and asks sequences of precise questions to the user. The user is then supposed to answer those questions and provide only the information he/she has been asked for.

User-led   the initiative is completely let to the user who asks for information to the system. The system is supposed to interpret correctly the user's query and to answer those precise questions without asking for more details.

Mixed initiative both participants (the user and the system) share the control to cooperate to achieve the conversation goal. The user can be over-informative and provide information he/she has not yet been asked for. He/she can ask the system to perform particular actions as well. The system can take the control at certain dialogue states so as not to deviate from the correct path leading to goal achievement. The dialogue manager can also decide to take the control because the performance of the previous systems in the chain (ASR, NLU) is likely to get poorer.

Instinctively, mixed-initiative systems are thought to potentially perform better from the user's point of view. However, some research has shown that users sometimes prefer system-led SDSs because the goal achievement rate is greater [65]. Other studies exemplified that, besides the fact that system-led gives better performance with inexperienced users, the mixed-initiative version of the same system does not outperform the first (objectively neither subjectively) with more experienced users [66]. Indeed, evidence is made that human users adapt their behaviour because they know they are interacting with a machine. They usually accept some constraints to obtain a better goal completion rate.

4.4.4 Evaluation

Previous sections underlined the main issues in dialogue management design such as the choices of a dialogue model, the degree of initiative, etc. In the literature, several studies lead to controversial conclusions about each of these issues. Moreover, objective evaluation criteria are used in machine learning methods for automatic optimisation of dialogue strategies (see Section 4.4.2). Consequently, SDS assessment methods became the topic of a large field of current research.

Despite the amount of research dedicated to the problem of SDS performance evaluation, there is no clear, objective and commonly accepted method to tackle this issue. Indeed, performance evaluation of such high-level communicative systems strongly relies on the opinion of end-users and is therefore strongly subjective and most probably task-dependent. Studies on subjective evaluation of SDSs through user satisfaction surveys have often (and early) been conducted [67]. The reproductivity of experiments and the independence between experiments is hardly achieved. This is why those experiments have non-trivial interpretation. For instance, in [68] the authors demonstrate as a side-conclusion that the users' appreciation of different strategies depends on the order in which SDSs implementing those strategies were presented for evaluation. Nevertheless, several attempts have been made to determine the overall system performance, thanks to objective measures made on the sub-components, such as ASR performance. One of the first tries can be found in [69]. Other objective measures taking into account the whole system behaviour (like the average number of turns per transaction, the task success rate, etc.) have been exercised in the aim of evaluating different versions (strategies) of the same SDS [70] with one of the first applications in the SUNDIAL project (and after within the EAGLES framework) [71]. More complex paradigms have been developed afterwards as described hereafter.

PARADISE

A popular framework for SDS evaluation is PARADISE (PARAdigm for DIalogue Systems Evaluation) [72]. This paradigm attempts to explain users' satisfaction as a linear combination of objective measures. For the purpose of evaluation, the task is described as an attribute-value matrix. The user's satisfaction is then expressed as the combination of a task completion measure (κ) and a dialogue cost computed as a weighted sum of objective measures (c_i). The overall system performance is then approximated by:

$$P(U) = \alpha N(\kappa) + \sum_i w_i N(c_i) \qquad (4.8)$$

where N is a Z-score normalisation function that normalises the results to have mean 0 and standard deviation 1. This way, each weight (α and w_i) expresses the relative contribution of each term of the sum to the performance of the system. The task completion measure κ is the Kappa coefficient [73] that is obtained from a confusion matrix M summarising how well the transfer of information performed between the user and the system. M is a square matrix of dimension n (number of values in the attribute-value matrix) where each element m_{ij} is the number of dialogues in which the value i was interpreted while value j was meant. The κ coefficient is then defined as:

$$\kappa = \frac{P(A) - P(E)}{1 - P(E)} \tag{4.9}$$

where $P(A)$ is the proportion of correct interpretations (sum of the diagonal elements of M (m_{ii}) on the total number of dialogues) and $P(E)$ is the proportion of correct interpretations occurring by chance. One can see that $\kappa = 1$ when the system performs perfect interpretation ($P(A) = 1$) and $\kappa = 0$ when the only correct interpretations were obtained by chance ($P(A) = P(E)$). So as to compute optimal weights α and w_i, test users are asked to answer a satisfaction survey after having used the system while costs c_i are measured during the interaction. The questionnaire includes around nine statements to be rated on a five-point Likert scale and the overall satisfaction is computed as the mean value of collected ratings. A multivariate linear regression (MLR) is then applied with the result of the survey as the dependent variable and the weights as independent variables. Criticisms can be made about assumptions and methods involved in the PARADISE framework. First, the assumption of independency between the different costs c_i made when building an additive evaluation function has never been proven (it is actually false as the number of turns and the time duration of a dialogue session are strongly correlated [74]). The κ coefficient as a measure of task success rate can also be discussed, as it is often very difficult to compute when a large number of values are possible for a given attribute. In [75] this is exemplified by the application of PARADISE to the PADIS system (Philips Automatic Directory Information System). The satisfaction questionnaire has also been the subject of criticisms. Although [76] proposes to add a single statement rating the overall performance of the system on a 10-point scale, [74] recommends to rebuild the whole questionnaire, taking psychometric factors into account. Finally, the attribute-value matrix representation of the task has proven to be hard to extend to multimodal systems and thus, seems not to be optimal for system comparisons. Some attempts to modify PARADISE have been proposed to evaluate multimodal interfaces [77]. Besides the aforementioned criticisms, PARADISE has the advantage of being reproducible and involves automatic computations. It has thus been applied on a wide range of systems. It was adopted as the evaluation framework for the DARPA Communicator project and applied to the official 2000 and 2001 evaluation experiments [78]. However, experiments on different SDSs reached different conclusions. PARADISE developers themselves found contradicting results and reported that time duration was weakly correlated with user satisfaction in [79] while [80] reports that dialogue duration, task success and ASR performance were good predictors of user's satisfaction. On the other hand, [81] reports a negative correlation between users' satisfaction and dialogue duration because users hung up when unsatisfied which reduces the dialogue length. Finally, [82] surprisingly reports that ASR performance is not so good a predictor of user's satisfaction.

Analytical Evaluation

The principal drawback of PARADISE is the need of data collection. Indeed, subjective evaluation means that the system should be released to be evaluated and that a large number of test users has to be found. It is a time-consuming and expensive process that has to be done again for each prototype release. Although the usual process of SDS design obeys to the classical prototyping cycle composed of successive pure design and user evaluation cycles, there should be as little user evaluations as possible. This is why attempts to analyse strategies by mathematical means have been developed. In [83], the authors propose some ways to diagnose the future performance of the system during the design process. Other mathematical models of dialogue have been proposed [84] and closed forms of dialogue metrics (like the number of dialogue turns) have been discussed. Nevertheless, too few implementations were made to prove their reliability. Moreover, several simplifying assumptions have to be made for analytical evaluation and it is thus difficult to extend those methods to complex dialogue configurations and task-independency is hardly reachable.

Computer-Based Simulation

Because of inherent difficulties of data collection, some efforts have been done in the field of dialogue simulation for performance assessment. The purpose of using dialogue simulation for SDS evaluation is mainly to enlarge the set of available data and to predict the behaviour of the SDS in unseen situations. Among simulation methods presented in the literature, one can distinguish between state-transition methods as proposed in [54] and methods based on modular simulation environments as described in [48, 58–60]. The first type of method is more task-dependent as well as the hybrid method proposed in [85]. One can also distinguish methods according to the level of abstraction in which the simulation takes place. While [59, 86] models the dialog at the acoustic level, most of other methods [48, 54, 58, 60, 85] remain at the intention level, arguing that simulation of other levels can be inferred from intentions.

4.5 CONCLUSION

In this chapter, we have described processing systems that are usually hidden to the user although essential for building speech- or text-based interfaces. All of these systems are still being the topic of intensive research and there exists room for improvement in performance. Especially, data-driven methods for optimising end-to-end systems from speech recognition to text-to-speech synthesis are being investigated [87], albeit data collection and annotation is still a major problem. What is more, language processing is still limited to domain-dependent applications (such as troubleshooting, database access, etc) and cross-domain or even cross-language methods are still far from being available. Also, transfer of academic research into the industrial world is still rare [88]. The search for efficiency often leads to hand-crafted and system-directed management strategies that are easier to understand and control.

REFERENCES

1. J. Allen, Natural Language Understanding, second ed., Benjamin Cummings, 1994.
2. N. Chomsky, Three models for description of languages, Trans. Inf. Theory, 2 (1956) 113–124.
3. A. Aho, J. Ullman, The Theory of Parsing, Translation, and Compiling, Prentice-Hall, 1972.
4. W. Woods, Transition network grammars for natural language analysis, Commun. ACM, 13 (1970) 591–606.

5. D.E. Knuth, Backus normal form vs. backus naur form. Commun. ACM 7 (12) (1964) 735–736.

6. G. Gazdar, C. Mellish, Natural Language Programming in PROLOG, Addison-Wesley, Reading, MA, 1989.

7. F. Jelinek, Self-organized language modelling for speech recognition, in: A. Waibel, K.-F. Lee (Eds.), Readings in Speech Recognition, Morgan Kaufmann, 1990, pp. 450–506.

8. S. Seneff, Tina: a natural language system for spoken language applications, Comput. Linguist. 18 (1) (1992) 61–86.

9. J. Austin, How to Do Things with Words, Harvard University Press, Cambridge, MA, 1962.

10. P. Cohen, C. Perrault, Elements of a plan-based theory of speech acts, Cogn. Sci. 3 (1979) 117–212.

11. R. Montague, Formal Philosophy, Yale University, New Haven, 1974.

12. R. Pieraccini, E. Levin, Stochastic representation of semantic structure for speech understanding, Speech Commun. 11 (1992) 238–288.

13. Y. He, S. Young, Spoken language understanding using the hidden vector state model, Speech Commun. 48 (3–4) (2006) 262–275.

14. S. Pradhan, W. Ward, K. Hacioglu, J.H. Martin, D. Jurafsky, Shallow semantic parsing using support vector machines, in: Proceedings of HLT-NAACL, 2004.

15. C. Raymond, G. Riccardi, Generative and discriminative algorithms for spoken language understanding, in: Proceedings of Interspeech, Anvers (Belgium), August 2007.

16. F. Mairesse, M. Gašić, F. Jurčíček, S. Keizer, B. Thomson, K. Yu, et al., Spoken language understanding from unaligned data using discriminative classification models, in: Proceedings of ICASSP, 2009.

17. F. Jelinek, J. Lafferty, D. Magerman, R. Mercer, A. Ratnaparkhi, S. Roukos, Decision tree parsing using a hidden derivation model, in: HLT '94: Proceedings of the workshop on Human Language Technology, Morristown, NJ, USA, 1994, pp. 272–277, Association for Computational Linguistics.

18. L. Kartunnen, Discourse referents, in: J. McCawley (Ed.), Syntax and Semantics 7, Academic Press, 1976, pp. 363–385.

19. C. Sidner, Focusing in the comprehension of definite anaphora, in: M. Brody, R. Berwick (Eds.), Computational Models of Discourse, MIT Press, Cambridge, Mass, 1983, pp. 267–330.

20. M. Walker, A. Joshi, E. Prince (Eds.), Centering Theory in Discourse, Oxford University Press, 1998.

21. E. Reiter, R. Dale, Building Natural Language Generation Systems, Cambridge University Press, Cambridge, 2000.

22. B. Lavoie, O. Rambow, E. Reiter, Customizable descriptions of object-oriented models, in: Proceedings of the Conference on Applied Natural Language Processing (ANLP'97), Washington, DC, pp. 253–256, 1997.

23. E. Goldberg, N. Driedger, R. Kittredge, Using natural-language processing to produce weather forecasts, IEEE Expert Intell. Syst. Appl. 9 (2) (1994) 45–53.

24. O. Rambow, S. Bangalore, M. Walker, Natural language generation in dialog systems, in: Proceedings of the 1st International Conference on Human Language Technology Research (HLT'01), San Diego, USA, pp. 1–4, 2001.

25. R. Higashinaka, M. Walker, R. Prasad, An unsupervised method for learning generation dictionaries for spoken dialogue systems by mining user reviews, J. ACM Trans. Speech Lang. Processing (2007) 4 (4), 8.

26. S. Janarthanam, O. Lemon, Learning lexical alignment policies for generating referring expressions for spoken dialogue systems, in: Proceedings of ENLG, pp. 74–81, 2009.

27. M. Walker, J. Cahn, S. Whittaker, Linguistic style improvisation for lifelike computer characters, in: Proceedings of the AAAI Workshop AI, Alife and Entertainment, Portland, 1996.

28. S. Shieber, G. van Noord, F. Pereira, R. Moore, Semantic head-driven generation, Comput. Linguist. 16 (1990) 30–42.

29. S. McRoy, S. Channarukul, S. Ali, Text realization for dialog, in: Proceedings of the International Conference on Intelligent Technologies, Bangkok (Thailand), 2000.

30. M. Walker, O. Rambow, M. Rogati, Training a sentence planner for spoken dialogue using boosting, Comput. Speech Lang. Spec. Issue Spoken Lang. Generation, 16 (3–4) (2002) 409–433.

31. A. Jonsson, N. Dahlback, Talking to a computer is not like talking to your best friend, in: Proceedings of the 1st Scandinivian Conference on Artificial Intelligence, Tromso, Norway, 1988, pp. 53–68.

32. N. Dahlback, A. Jonsson, An empirically based computationally tractable dialogue model, in: Proceedings of the 14th Annual Conference of the Cognitive Science Society (COGSCI'92), Bloomington, Indiana, USA, pp. 785–791, 1992.

33. R. Reichman, Plain-Speaking: A Theory and Grammar of Spontaneous Discourse, PhD thesis, Department of Computer Science, Harvard University, Cambridge, Massashussetts, 1981.

34. S. Carberry, ACL-MIT Press Series in Natural Language Processing, chapter Plan Recognition in Natural Language Dialogue, Bradford Books, MIT Press, 1990.

35. T. Bylander, Complexity results for planning, in: Proceedings of the 12th International Joint Conference on Artificial Intelligence (IJCAI'91), 1991, pp. 274–279.

36. R. Freedman, Plan-based dialogue management in a physics tutor, in: Proceedings of the 6th Applied Natural Language Processing Conference, Seattle, WA, USA, 2000, pp. 52–59.
37. R. Power, The organization of purposeful dialogues, Linguistics, 17 (1979) 107–152.
38. G. Houghton, S. Isard, Why to speak, what to say, and how to say it, in: P. Morris (Ed.), Modelling Cognition, Wiley, 1987, pp. 249–267.
39. I. Lewin, A formal model of conversational games theory, in: Proceedings of the 4th Workshop on the Semantics and Pragmatics of Dialogues, GOTALOG'00, Gothenburg, 2000.
40. H. Clarck, E. Schaefer, Contributing to discourse, Cogn. Sci. 13 (1989) 259–294.
41. P. Edmonds, A computational model of collaboration on reference in direction-giving dialogues, Master's thesis, Computer Systems Research Institute, Department of Computer Science, University of Toronto, 1993.
42. R. Smith, R. Hipp, Spoken Natural Language Dialog Systems: a Practical Approach, Oxford University Press, New York, 1994.
43. M. McTear, Modelling spoken dialogues with state transition diagrams: Experiences with the CSLU toolkit, in: Proceedings of the 5th International Conference on Spoken Language Processing (ICSLP'98), Sydney (Australia), 1998.
44. W3C, VoiceXML 3.0 Specifications, December 2008, <http://www.w3.org/TR/voicexml30/>.
45. D. Goddeau, H. Meng, J. Polifroni, S. Sene, S. Busayapongchai, A form-based dialogue manager for spoken language applications, in: Proceedings of the 4th International Conference on Spoken Language Processing (ICSLP'96), Philadelphia, PA, USA, 1996, pp. 701–704.
46. H. Aust, O. Schroer, An overview of the Philips dialog system, in: Proceedings of the DARPA Broadcast News Transcription and Understanding Workshop, Lansdowne, Virginia, USA, 1998.
47. A. Baekgaard, Dialogue management in a generic dialogue systems, in: Proceedings of TWLT-11: Dialogue Management in Natural Language Systems, Netherlands, 1996, pp. 123–132.
48. O. Pietquin, T. Dutoit, A probabilistic framework for dialog simulation and optimal strategy learning, IEEE Trans. Audio Speech Lang. Process. 14 (2) (2006) 589–599.
49. A. Graesser, K. VanLehn, C. Rosé, P. Jordan, D. Harter, Intelligent tutoring systems with conversational dialogue, AI Mag. 22 (4) (2001) 39–52.
50. O. Pietquin, Machine learning for spoken dialogue management: an experiment with speech-based database querying, in: J. Euzenat, J. Domingue (Eds.), Artificial Intelligence: Methodology, Systems and Applications, vol. 4183 of Lecture Notes in Artificial Intelligence, Springer Verlag, 2006, pp. 172–180.

51. O. Pietquin, A Framework for Unsupervised Learning of Dialogue Strate-
    gies, SIMILAR Collection, Presses Universitaires de Louvain, 2004, ISBN:
    2-930344-63-6.
52. R. Sutton, A. Barto, Reinforcement Learning: An Introduction, MIT Press, 1998,
    ISBN: 0-262-19398-1.
53. E. Levin, R. Pieraccini, W. Eckert, Learning dialogue strategies within the
    Markov decision process framework, in: Proceedings of the International
    Workshop on Automatic Speech Recognition and Understanding (ASRU'97),
    pp. 72–79, 1997.
54. S. Singh, M. Kearns, D. Litman, M. Walker, Reinforcement learning for spoken
    dialogue systems, in: Proceedings of the Neural Information Processing Society
    Meeting (NIPS'99), Vancouver (Canada), 1999.
55. M. Frampton, O. Lemon, Learning more effective dialogue strategies using lim-
    ited dialogue move features, Proceedings of the 21st International Conference
    on Computational Linguistics and the 44th Annual Meeting of the Association
    for Computational Linguistics, pp. 185–192, 2006.
56. P.P.J. Williams, S. Young, Partially observable Markov decision processes
    with continuous observations for dialogue management, in: Proceedings of the
    SigDial Workshop (SigDial'06), 2005.
57. S. Young, Using POMDPS for dialog management, in: Proceedings of the 1st
    IEEE/ACL Workshop on Spoken Language Technologies (SLT'06), 2006.
58. E. Levin, R. Pieraccini, W. Eckert, A stochastic model of human–machine inter-
    action for learning dialog strategies, IEEE Trans. Speech Audio Process. 8 (1)
    (2000) 11–23.
59. R. López-Cózar, A. de la Torre, J. Segura, A. Rubio, Assesment of dialogue
    systems by means of a new simulation technique, Speech Commun. 40 (3)
    (2003) 387–407.
60. O. Pietquin, A probabilistic description of man–machine spoken communica-
    tion, in: Proceedings of the International Conference on Multimedia and Expo
    (ICME'05), Amsterdam (Netherlands), July 2005.
61. J. Schatzmann, K. Weilhammer, M. Stuttle, S. Young, A survey of statistical
    user simulation techniques for reinforcement-learning of dialogue management
    strategies, Knowl. Eng. Rev. 21 (2) (2006) 97–126.
62. O. Pietquin, T. Dutoit, Dynamic Bayesian networks for NLU simulation with
    application to dialog optimal strategy learning, in: Proceedings of the IEEE
    International Conference on Acoustics, Speech and Signal Processing (ICASSP
    2006), vol. 1 (May 2006), pp. 49–52, Toulouse (France).
63. O. Pietquin, S. Renals, ASR system modelling for automatic evaluation and
    optimization of dialogue systems, in: Proceedings of the IEEE International

Conference on Acoustics, Speech and Signal Processing (ICASSP 2002), vol. 1 (May 2002) pp. 45–48, Orlando, (USA, FL).

64. J. Schatzmann, B. Thomson, S. Young, Error simulation for training statistical dialogue systems, in: Proceedings of the International Workshop on Automatic Speech Recognition and Understanding (ASRU'07), Kyoto (Japan), 2007.

65. J. Potjer, A. Russel, L. Boves, E. den Os, Subjective and objective evaluation of two types of dialogues in a call assistance service, in: Proceedings of the IEEE Third Workshop on Interactive Voice Technology for Telecommunications Applications (IVTTA 96), 1996, pp. 89–92.

66. M. Walker, D. Hindle, J. Fromer, G.D. Fabbrizio, C. Mestel, Evaluating competing agent strategies for a voice email agent, in: Proceedings of the 5th European Conference on Speech Communication and Technology (Eurospeech'97), Rhodes (Greece), 1997.

67. J. Polifroni, L. Hirschman, S. Seneff, V. Zue, Experiments in evaluating interactive spoken language systems, in: Proceedings of the DARPA Speech and Natural Language Workshop, Harriman, NY (USA), February 1992, pp. 28–33.

68. L. Devillers, H. Bonneau-Maynard, Evaluation of dialog strategies for a tourist information retrieval system, in: Proceedings of the 5th International Conference on Speech and Language Processing (ICSLP'98), Sydney, Australia, 1998, pp. 1187–1190.

69. L. Hirschman, D. Dahl, D. McKay, L. Norton, M. Linebarger, Beyond class a: a proposal for automatic evaluation of discourse, in: Proceedings of the DARPA Speech and Natural Language Workshop, 1990, pp. 109–113.

70. M. Danieli, E. Gerbino, Metrics for evaluating dialogue strategies in a spoken language system, in: Working Notes of the AAAI Spring Symposium on Empirical Methods in Discourse Interpretation and Generation, Stanford, CA (USA), 1995, pp. 34–39.

71. N.F.A. Simpson, Black box and glass box evaluation of the sundial system, in: Proceedings of the 3rd European Conference on Speech Communication and Technology (Eurospeech'93), Berlin, Germany, 1993, pp. 1423–1426.

72. M. Walker, D. Litman, C. Kamm, A. Abella, Paradise: A framework for evaluating spoken dialogue agents, in: Proceedings of the 35th Annual Meeting of the Association for Computational Linguistics, Madrid (Spain), 1997, pp. 271–280.

73. J. Carletta, Assessing agreement on classification tasks: the kappa statistic, Comput. Linguist, 22 (2) (1996) 249–254.

74. L. Larsen, Issues in the evaluation of spoken dialogue systems using objective and subjective measures, in: Proceedings of the IEEE Automatic Speech Recognition and Understanding Workshop (ASRU'03), pp. 209–214, St. Thomas (U.S. Virgin Islands), 2003.

75. G. Bouwman, J. Hulstijn, Dialogue strategy redesign with reliability measures, in: Proceedings of the 1st International Conference on Language Resources and Evaluation, 1998, pp. 191–198.

76. P. Sneele, J. Waals, Evaluation of a speech-driven telephone information service using the paradise framework: a closer look at subjective measures, in: Proceedings of the 8th European Conference on Speech Technology (EuroSpeech'03), Geneva (Switzerland), September 2003, pp. 1949–1952.

77. N. Beringer, U. Kartal, K. Louka, F. Schiel, U. Türk, Promise – a procedure for multimodal interactive system evaluation, in: Proceedings of the Workshop on Multimodal Resources and Multimodal Systems Evaluation, pp. 77–80, Las Palmas, Gran Canaria, Spain, 2002.

78. M. Walker, L. Hirschman, J. Aberdeen, Evaluation for DARPA communicator spoken dialogue systems, in: Proceedings of the Language Resources and Evaluation Conference (LREC'00), 2000.

79. M. Walker, J. Boland, C. Kamm, The utility of elapsed time as a usability metric for spoken dialogue systems, in: Proceedings of the IEEE Automatic Speech Recognition and Understanding Workshop (ASRU'99), Keystone, CO (USA), 1999.

80. M. Walker, J. Aberdeen, J. Boland, E. Bratt, J. Garofolo, L. Hirschman, et al., DARPA communicator dialog travel planning systems: The June 2000 data collection, in: Proceedings of the 7th European Conference on Speech Technology (Eurospeech'01), Aalborg (Danmark), 2001.

81. M. Rahim, G.D. Fabbrizio, M.W.C. Kamm, A. Pokrovsky, P. Ruscitti, E. Levin, et al., Voice-if: a mixed-initiative spoken dialogue system for AT & T conference services, in: Proceedings of the 7th European Conference on Speech Processing (Eurospeech'01), Aalborg (Denmark), 2001.

82. L. Larsen, Combining objective and subjective data in evaluation of spoken dialogues, in: Proceedings of the ESCA Workshop on Interactive Dialogue in Multi-Modal Systems (IDS'99), Kloster Irsee, Germany, 1999, pp. 89–92.

83. D. Louloudis, K. Georgila, A. Tsopanoglou, N. Fakotakis, G. Kokkinakis, Efficient strategy and language modelling in human–machine dialogues, in: Proceedings of the 5th World Multi-Conference on Systemics, Cybernetics and Informatics (SCI'01), vol. 13 (2001), pp. 229–234, Orlando, Fl (USA).

84. Y. Niim, T. Nishimoto, Mathematical analysis of dialogue control strategies, in: Proceedings of the 6th European Conference on Speech Technology, (EuroSpeech'99), Budapest, September 1999.

85. K. Scheffler, S. Young, Corpus-based dialogue simulation for automatic strategy learning and evaluation, in: Proceedings of NAACL Workshop on Adaptation in Dialogue Systems, 2001.

86. R. López-Cózar, Z. Callejas, M.F. McTear, Testing the performance of spoken dialogue systems by means of an artificially simulated user, Artif. Intell. Rev. 26 (4) (2006) 291–323.

87. O. Lemon, O. Pietquin, Machine learning for spoken dialogue systems, in: Proceedings of the European Conference on Speech Communication and Technologies (Interspeech'07), Anvers, Belgium, 2007, pp. 2685–2688.

88. K. Acomb, J. Bloom, K. Dayanidhi, P. Hunter, P. Krogh, E. Levin, R. Pieraccini, Technical support dialog systems, issues, problems, and solutions, in: Proceedings of the HLT 2007 Workshop on "Bridging the Gap, Academic and Industrial Research in Dialog Technology," pp. 25–31, Rochester, NY, USA, 2007.

Chapter 5

Image and Video Processing Tools for HCI

Montse Pardàs, Verónica Vilaplana
and Cristian Canton-Ferrer
Universitat Politècnica de Catalunya, Barcelona, Spain

5.1 INTRODUCTION

Image and video processing tools for human–computer interaction (HCI) are reviewed in this chapter. Different tools are used in close

Multimodal Signal Processing, ISBN: 9780123748256

view applications, such as desktop computer applications or mobile telephone interfaces, and in distant view setups, such as smart-rooms scenarios or augmented reality games. In the first case, the user can be captured in a close view and some assumptions can be made regarding the location and pose of the user. For instance, in face-oriented applications, a frontal view of the face is generally assumed, whereas for gestural interfaces, the hand is supposed to perform a gesture from a specific dictionary directly in front of the camera. We will review in Sections 5.2 and 5.3 of this chapter the current state of the art for face and gesture analysis in video sequences in this context. To lighten these restrictions and to allow a wider set of applications, in the recent years a high interest has been developed for distant view scenarios, being smart-rooms the most complex paradigm. This kind of scenario typically contains multiple cameras distributed around the scene, with partial overlap among them, and thus a multiview analysis can be performed. The applications in this case can be designed with less restrictions and the full body of the user can be analysed. In Sections 5.4 and 5.5, we will describe two areas of research that have been studied both from the single camera approach and from the multicamera point of view.

The image processing techniques that we will describe will cover from the low-level analysis to the semantics. The pipeline used is common to most applications: detection (face, hand or person), feature extraction (where the features will be specifically designed for each application) and recognition (of the person, the expression or the gesture, for instance). In this chapter, we will review the state of the art of the lower level techniques and present examples of the more important HCI tools that use these algorithms.

5.2 FACE ANALYSIS

Face analysis is used in HCI for recognition of the person and for more advanced interfaces that take into account the user state, analysing for instance its facial expressions. In this section, we will first review the low-level analysis tools that are used for the detection (Section 5.2.1) and tracking (Section 5.2.2) of the face. The facial feature extraction and tracking problem is analysed in Section 5.2.3, whereas Section 5.2.4 tackles the problem of gaze analysis. Finally, we will overview the two main applications: face recognition (Section 5.2.5) and facial expression analysis (Section 5.2.6).

5.2.1 Face Detection

Face detection is a necessary first step in any face analysis system. Although it is a visual task that humans perform quite effortlessly, face detection is a challenging problem in image analysis and computer vision, which has to be solved regardless of clutter, occlusions, variations in illumination, head pose, facial expressions, etc.

Face detection, defined as the problem of finding the location and extension of all the human faces present in the scene, has been approached with a diversity of strategies (see surveys [1, 2]). The techniques can be broadly categorised into block-based, region-based and feature-based methods, taking into account the model used to represent the patterns in the image.

Block-based methods: They define rectangular candidate areas by scanning the image (typically the luminance component) with a sliding window of fixed size. To detect faces of different sizes, the image is repeatedly down-scaled, and the analysis is performed at each scale. Next, a set of local or global features is extracted for each rectangle and used to classify it into face or nonface. Various classifiers can be used for this purpose: neural networks [3], support vector machines (SVM) [4], Sparse network of windows [5], naive Bayes classifiers [6], Adaboost [7], etc.

To reduce the computational cost of the search over an exhaustive range of locations and scales in the image, the most recent techniques use a coarse to fine cascade of classifiers of increasing complexity [8]. At each stage, a classifier decides either to reject the pattern or to continue the evaluation using the next classifier. This strategy is designed to remove most of the candidates with a minimum amount of computation. Most of the image is quickly rejected as nonface, and more complex classifiers concentrate on the most difficult face-like patterns.

Region-based methods: They work with a small number of candidate regions which are typically defined using colour information. Some of these techniques use colour as a first cue and classify image pixels into skin or nonskin coloured; skin-like pixels are then grouped and evaluated by relatively simple classifiers [9, 10]. Other approaches partition the image into regions homogeneous in colour and create a hierarchy of regions representing the image at different resolution levels. Next, regions are analysed by a combination of one-class classifiers based on visual attributes of faces [11].

Feature-based methods: They search for facial features: eyebrows, eyes, nose, mouth, hair-line or face contours. Then, anthropometric measures or statistical models that describe the relationship between features are used to verify the candidate configurations [12, 13]. Techniques for facial feature detection are detailed in Section 5.2.3.

The current state of the art on face detection is based on the real-time block-based method proposed by Viola and Jones [7], which combines the idea of a cascade of classifiers with the efficient computation of local features and a learning algorithm based on Adaboost. The algorithm works with an intermediate representation of the image, the so-called *integral image*, that allows a fast computation of a set of local 'Haar-like' features. Four types of rectangle features are used, which compute differences between the sums of pixels within rectangular areas, giving over 160.000 features for patterns of size 24×24. Weak classifiers, which are decision stumps based on these local features, are boosted into a single strong classifier. At each step of the training, the weak classifier with minimum error rate on the training set is selected. In turn, strong classifiers are combined in a coarse to fine cascade. Each classifier in the cascade is trained using negative samples, that is, false positives output by the previous classifier. The approach was extended in [14] with an additional set of 45° rotated Haar-like features and improved using GentleBoost to train the classifiers [15].

5.2.2 Face Tracking

Once a face has been detected, it is often necessary to track its position. There are three main approaches to the problem: frame-based detection, where the face is detected in each frame without taking into account the temporal information, separated detection and tracking, in which the face is detected in the first frame and then is tracked in the following frames, and integrated approaches, where temporal relationships between frames are used to detect the face through the sequence. Although the first approach can rely on any of the face detection strategies previously discussed, the other two approaches typically apply techniques specifically developed for the more general problem of object tracking.

Object tracking is a complex problem due to several factors: the loss of information caused by the projection from a 3D real scene to a 2D image, image noise, the nonrigid or articulated nature of objects, complex object motion, partial or full occlusions, illumination changes and real-time processing requirements.

In a recent survey, Yilmaz et al. [16] organised tracking methods (for both separated detection and tracking and integrated approaches) in three main groups, taking into account the model used to represent the object shape: point, kernel and silhouette-based methods.

Point tracking methods represent objects by points (e.g., the object centroid or a set of characteristic points). The tracker establishes a correspondence between points in consecutive frames. Point correspondence methods can be deterministic or statistical. Deterministic methods define a correspondence cost of associating points in consecutive frames taking into account motion, proximity or velocity constraints, and the cost is minimised, formulating the problem as a combinatorial optimisation problem. Statistical methods, in turn, take into account measurement and model uncertainties and use a state space approach to model object properties like position, velocity or acceleration. Kalman filters, particle filters and joint probabilistic data association filters belong to this class. Face tracking systems based on this type of tracker are deterministic [17] and statistical [18, 19].

In kernel tracking, the term kernel refers to the object shape or appearance (e.g., a rectangular or an elliptical template or an appearance model [20] inside a shape). The objects are tracked by estimating the motion, typically a parametric transformation, of the kernel in consecutive frames. The object motion is estimated by maximising the object appearance similarity between the previous and the current frame. Some face trackers based on these strategies are proposed in [20–22].

Finally, silhouette-based techniques represent the object by a region (an image mask) or by contours (a set of control points or level sets, for instance). The tracking is performed by estimating the area of support of the object in each frame, using for instance shape matching techniques for the case of region representations and methods that evolve an initial contour in the previous frame to its new position in the current frame, for contour representations. Face tracking systems in this group are proposed in [23, 24].

5.2.3   Facial Feature Detection and Tracking

Automatic facial feature (eyes, mouth, eyebrows, face contour) detection and tracking within the detected face area is used for different applications in human–computer interfaces: face recognition, emotion recognition, focus of attention (FoA) tracking, etc. The difficulty of this task, as in the case of face detection, is a consequence of the large variations that a face can show in a scene due to several factors: position, facial orientation, expression, pose and illumination.

Although most facial feature detection and tracking systems rely on a model to some extent, we can distinguish between systems that perform a local search, model-based systems and hybrid systems.

Local feature detection: In these methods, a pattern for each specific feature of interest is used for the search. For instance, in [25], the detected face region is divided into 20 relevant regions, each of which is examined further to predict the location of the facial feature points. The facial feature point detection method uses individual feature patch templates to detect points in the relevant region of interest. These feature models are GentleBoost templates built from both grey level intensities and Gabor wavelet features. Another example is [26], where 1D horizontal and vertical projections of edge images are used to obtain candidates for the eyes positions. Those candidates are filtered using SVM trained with eye patterns. Those systems currently use point tracking algorithms for feature tracking. Valstar and Pantic [25] track the facial points with a particle filtering scheme with factorised likelihoods and an observation model that combines a rigid and a morphological model.

Global model detection: More robust techniques for detection and tracking can be constructed using deformable 2D or 3D face models. The face model constrains the solutions to be valid examples of the object modelled. The first works in this direction [27] only used the model for tracking purposes, and the initialisation of the facial features was done manually. The model was defined as a nonuniform mesh of polyhedral elements, and the search was conducted with an extension of active contours (snakes). More recently, elastic bunch graph matching (EBGM) [28] has been proposed as a model-based system that does not need manual initialisation, once the face bunch graph (FGB) has been constructed. A first set of graphs is generated manually, locating nodes at fiducial points. Edges between the nodes as well as correspondences between nodes of different poses are defined. Once the system has an FBG, graphs for new images can be generated automatically by EBGM. Matching an FBG on a new image is done by maximising a graph similarity between an image graph and the FBG.

A very popular method is the active appearance model (AAM) algorithm due to Cootes et al. [20]. It is an analysis-by-synthesis system that uses photo-realistic models which are matched directly by minimising the difference between the image under interpretation and one synthesised by the model. It was developed as an extension of the active shape model, which models grey-level texture using local linear template models and the configuration of feature points using a statistical shape model. An iterative search algorithm seeks to improve the match of local feature points and then refines the feature locations by the best fit of the shape model. Fitting to a large number of facial feature points mitigates the effect of individual erroneous feature detections. The combination of shape and texture leads to the AAM approach to model matching. The shape and texture are combined in one principal component analysis (PCA) space. The model then iteratively searches a new image by using the texture residual to drive the model parameters. An example of facial feature detection with this method is shown in Figure 5.1.

Hybrid systems: Although model-based systems are more robust than local-based feature detectors, they are prone to fall in local

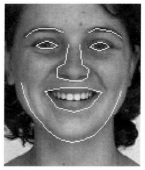

FIGURE 5.1   Example of facial feature detection. Face has been detected with Viola and Jones [7] and facial features with the AAM software [29].

minima, in particular for the initialisation. Thus, some authors perform it in a coarse-to-fine manner, including them in multiscale frameworks, and combine them with local feature detectors. In [30], Gabor filters are used for individual feature detection, and triplets of appropriate configurations of features are tested using an SVM model of facial appearance. The AAM in [31] applies individual feature detectors that are then combined using an algorithm known as pairwise reinforcement of feature responses. The points predicted by this method are then refined using an AAM search. Finally, some systems use a facial feature detector to initialise a 3D model in the first frame and then perform the tracking using this model. For instance, [32] uses in this framework an explicit 3D wireframe face model. The model, consisting of 16 surface patches embedded in Bezier volumes, is warped to fit the selected facial features.

5.2.4   Gaze Analysis

Gaze, as a strong indicator of attention, has been extensively studied in psychology and more recently in neuroscience and HCI. An eye gaze tracker is a system that estimates the direction of gaze of a person [33]. Early systems, developed mainly for diagnosis purposes, are very intrusive techniques, but the most accurate: contact lenses with magnetic field sensors or electrodes that measure differences in skin potentials around the eye. Other intrusive methods include cameras or other optical devices that need to be very close to the eye and are, therefore, head mounted or use a chin rest or a bite bar to restrict the head motion. Nonintrusive methods or remote eye gaze trackers (REGTs), in turn, are vision-based systems that use cameras to capture images of the eye and rely on eye properties that can be detected and tracked.

An important group of REGTs is infrared (IR)-based systems that work with IR or near-IR light sources, which are (almost) invisible for the human eye. If the IR source is placed near the optical axis of the camera, the light reflects off the retina and creates a bright pupil effect similar to the 'red eye' effect in photography with flash light. This effect creates a high contrast between iris and pupil and allows for robust pupil detection and tracking. The IR light also produces reflections at different layers of the eye, called the Purkinje images. The first Purkinje image or corneal reflection (CR), which is the reflection of the external corneal surface, is the easiest to detect and track, and it is therefore used by several systems. The position of the CR and the centre of the pupil define a vector in the image that can be used to find the direction of sight [34, 35]. To estimate the gaze direction, a calibration procedure is required. Typically, the subject looks at a set of points, while the system records the values corresponding to each gaze position. From these measurements, a calibration function is computed.

Other groups of REGTs are non-IR, appearance-based methods, which work with intensity images using computer vision techniques to detect the eyes and estimate their orientation. Many of these techniques can successfully locate eye regions [35, 36]. However, their accuracy in tracking eye movements is significantly lower than IR-based techniques.

5.2.5 Face Recognition

The problem of face recognition has received much attention over the last years; however, it is still very challenging in real applications because variations in pose or illumination may drastically change the appearance of a face [37]. Face recognition methods can be mainly categorised in appearance-based or model-based, according to the face representation they use.

Appearance-based methods: They use the whole face region as input to the recognition system. These methods represent face images as points in a high-dimensional vector space and use statistical techniques to analyse the distribution of these points in the space and derive an efficient representation (a 'feature space'). The most widely applied techniques use linear approaches such as PCA, independent component analysis or linear discriminant analysis (LDA). The *eigenface* method [38] assumes that face data lie in a low-dimensional manifold and uses PCA to reduce the dimensionality of the feature space and represent each face as a linear combination of a set of eigenvectors with largest eigenvalues. Recognition is performed by projecting the test image on the subspace defined by these eigenvectors and comparing the weights with those of the stored images. Several extensions have been proposed, such as modular eigenspaces [39] or probabilistic subspaces [40].

The *Fisherface* algorithm [41], based on LDA, exploits class-specific information, that is, intraclass and interclass variations between training samples, to find the vectors in the underlying space that best discriminate among classes. This technique results in a higher accuracy rate than the eigenfaces method but requires a sufficient number of samples per subject (class), which may not be available in real applications. Nonlinear manifold modelling algorithms are also used, such as kernel-PCA [42] or the Isomap algorithm [43].

Model-based methods: They work with a model of the human face, which is usually built using prior knowledge of faces. One of these techniques is EBGM [28], where faces are represented as graphs with nodes positioned at fiducial points, and edges labelled with distances between points. Each node contains a set of local descriptors, the wavelet coefficients extracted using different Gabor kernels. The graphs of individual sample faces are stacked to define a face class (FBG). For recognition, the FBG is positioned on the face image (see subsection 5.2.3) and then the face is identified by comparing the similarity between that face and all the faces stored in the FBG.

AAMs [20], already described in Section 5.2.3, are also used in face recognition. For all the training images, the model parameter vectors are used as feature vectors, and LDA is applied to build a discriminant subspace. Given a query image, AAM fitting is used to extract the corresponding feature vector. The recognition is achieved by finding the best match between query and stored feature vectors.

The aforementioned techniques referred to still images. Face recognition from video has gained interest during the last years for surveillance or monitoring applications. The main challenges of video-based face recognition are usually low quality of images and the small size and large variations in pose of faces. Many systems extend the techniques developed for still images, using video to select the best frames (e.g., frontal, well-illuminated faces) and combining individual recognition results through voting or other mechanisms [37]. Other techniques combine spatial and temporal information, for example, through a time series state space model [44].

5.2.6   Facial Expression Recognition

Several surveys on emotions recognition have been written in the last years [45–48]. Both video and audio analysis tools are used in this field. Concerning video analysis, most works develop a video database from subjects making facial expressions on demand. Their aim has been to classify the six basic emotions that psychological studies have indicated that are universally associated with distinct facial expressions: happiness, sadness, surprise, fear, anger and disgust [49]. Recently, some works have started to focus on authentic facial expressions in a more natural environment [50, 51]. Most approaches in this area rely on facial action coding system (FACS) [52]. The FACS is based on the enumeration of all action units (AUs) of a face that cause facial movements. The combination of these action units results in a large set of possible facial expressions. Another common approach is to use the MPEG4 standard parameters for facial animation (FAPs) that can be computed using a subset of the face definition parameters.

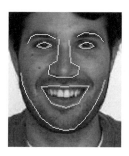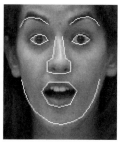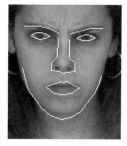

FIGURE 5.2　Facial features extracted for different facial expressions.

Whether using AUs or FAPs, most works rely on the results of facial feature detection and tracking that have been described in Section 5.2.3. We can see in Figure 5.2 the facial features extracted with an AAM for different facial expressions. Thus, facial expression recognition consists in the classification of the facial features movements using either spatial- or spatio-temporal classifiers [47]. Spatio-temporal classifiers commonly use hidden Markov models [53, 54], whereas spatial classifiers typically use neural networks. For instance, in [55], a radial basis function network architecture is developed that learns the correlation between facial feature motion patterns and human emotions. In [56], a dictionary is constructed to describe the facial actions of the FACS through the motion of the features, and a rule-based system is used to recognise facial expressions from these facial actions.

5.3　HAND-GESTURE ANALYSIS

Computer vision techniques for human motion capture (HMC) have been widely explored in the recent years [57]. Although techniques using markers or special hardware currently provide the most reliable results, they are not acceptable in many applications. An important field of research in HMC is the multicamera scenario, where 3D articulated human body models are used to track with precision the limb motion [58]. This scenario will be described in Section 5.5 of this chapter. In close view scenarios, such us desktop computer applications, the use of the hand as input device provides natural HCI. One approach is to use glove-based sensing or touch screens. However, computer vision-based approaches introduce less restrictions into the application and are more natural. A recent review of the computer vision-based research approaches to hand gesture recognition is provided in [59, 60].

Analogously to face or facial gesture recognition, the problem is usually decomposed in three successive steps: hand detection, feature detection and tracking, and gesture recognition.

The hand detection problem is often treated as background subtraction. Depending on the application, the problem can be simplified by using a zenital camera and an homogeneous background (the interactive surface) [61]. However, in most applications that require vertical gestures, the motion of the body of the person produces artifacts in common background subtraction techniques. Skin colour-based segmentation also presents problems in this context [62], due to the fact that the face of the person is also visible and corresponds to the same colour as the hand. A recent approach is to use time-of-flight range cameras that supply depth information per pixel [63]. Depth information can separate the object from the background much better than intensity images where colours, lighting, reflections and shadows have effect on almost every normal scenario [64]. Regarding hand feature detection and tracking two approaches are possible. The first one consists in using a 2D appearance model of the hand. A more complex and robust approach is to use an articulated 3D hand model.

Appearance-based methods: The detected hand is analysed with image processing techniques to extract the relevant features (usually the hand position and orientation, and the fingertips position of the extended fingers, or contour descriptors like Hu moments or Fourier descriptors [65]). A limited vocabulary of gestures must be defined that allows a mapping between the modelled hand and the set of gestures used. Both static and dynamic gestures are possible. The classification is usually performed using machine learning techniques (neural networks, Hidden Markov Models (HMM), etc.).

Model-based analysis: Model-based analysis uses tools developed for HMC with articulated body models that will be covered in Section 5.5. In this case, the model is limited to the hand, and the main difficulty comes from the fact that self-occlusions are strong and the motion is usually very fast. These techniques usually require a multicamera environment or at least one stereo camera, and are computationally more expensive. They are based on an analysis-by-synthesis scheme. That is, a search on the model parameters is performed looking for a minimisation of the error between the model configuration with a given set of parameters and the observed scene.

5.4 HEAD ORIENTATION ANALYSIS AND FoA ESTIMATION

Although face detection is enough for simple HCI, more advanced interfaces that aim at a higher semantic analysis of the scene require also an estimation of the head orientation. These applications usually do not restrict the position of the person, and thus, close views are not always available. We will review in this section how this estimation can be performed with a single camera and how multiple cameras can increase the robustness and alleviate the restrictions of the systems. From a mathematical viewpoint, multiple view formulation is usually addressed through epipolar geometry [66], and it allows exploiting the spatial redundancy among cameras towards circumventing effects produced by perspective or occlusions among objects present in the scene.

5.4.1 Head Orientation Analysis

Head orientation estimation can produce valuable information to estimate FoA, and a number of techniques have been devoted to this aim [67]. The general approach involves estimating the position of specific facial features in the image (typically eyes, nostrils and mouth) and then fitting these data to a head model. The accuracy and reliability of the feature extraction process play an important role in the head pose estimation results. However, facial feature detection and tracking, as reviewed in Section 5.2.3, currently assume that near-frontal views and high-quality images are available. For the applications addressed in a multicamera environment, such conditions are usually difficult to satisfy because specific facial features are not clearly visible due to lighting conditions and wide angle camera views. They may also be entirely unavailable when faces are not oriented towards the cameras. Furthermore, most of the existing approaches to head orientation analysis are based on monocular analysis of images, but few have addressed the multiocular case for face or head analysis [68].

Head orientation estimation using multiple cameras has been addressed as an estimation or classification problem. In the first case, it is required to estimate the head orientation angle using a prediction and update procedure within a Bayesian context. Techniques based on the particle filtering tracking algorithm [69] have proved efficient as shown in [70] where information provided by multiple views is used to generate a synthetic head representation and, afterwards, estimate its orientation. In the case where annotated data are available (as in the CLEAR head orientation dataset [71]), pattern recognition algorithms may be used to train a multiclass classifier where every class corresponds to a given orientation. Several classification approaches are proposed ranging from neural networks [72] to submanifold analysis [73]. However, when applying any of the aforementioned algorithms, it is required to know the position of the head beforehand. For this aim, face detection in multiview scenarios [74] or person tracking algorithms [75] can be used to locate the position of people in the scene and estimate the approximate location of the head.

5.4.2 Focus of Attention Estimation

Estimating the FoA from video inputs is a research topic involving computer vision techniques and psychology. In our context, FoA is understood as the region/person that captures the most of the attention of a person or a group of people (see Figure 5.3). Some higher semantic level applications can be derived from this information such as automatic meeting indexing [76] or capturing interactions among a group of people [75]. There are two main paths to infer FoA based on the fact that the attention of a person is usually aligned with the direction where he or she looks at [77], and therefore head orientation estimation and eye tracking can be used to assess FoA. Although this last technique may procude very accurate results regarding the spatial region where the person is looking at, intrusive and/or specialised hardware is usually required, as described in Section 5.2.4.

Extracting FoA information from head orientation has been typically addressed in the literature using neural networks [76] or

FIGURE 5.3 Focus of attention estimation of a group of people may allow the system to identify this gathering as a lecture where somebody is distracted looking through the window and somebody else is checking his email at the computer.

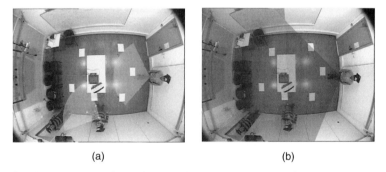

(a)   (b)

FIGURE 5.4 Focus of attention descriptors. In (a), the attention cones generated by two people. In (b), the attention map generated by the intersection of two cones: blue denotes the areas with one intersection and red the areas with two intersections.

HHMs [78]. The spatial region where the attention of a person is drawn is tightly correlated with the orientation of his head and the horizontal and vertical span of his eyes perception [77]. Other approaches aim at keeping an affordable complexity of the system and introduce geometric descriptors towards estimating the FoA [75] such as the attention cones (see Figure 5.4).

5.5   BODY GESTURE ANALYSIS

Analysis of human motion and gesture in image sequences is a topic that has been studied extensively [79]. Detection and recognition of several human centred actions are the basis of these studies. This analysis is important to build attentive interfaces aiming at supporting humans in various tasks and situations. Examples of these intelligent environments include the 'digital office' [80], 'intelligent house', 'intelligent classroom' and 'smart conferencing rooms' [81]. When addressing this problem, it is usually required to first extract some parameters relative to the motion of people in the scene and analyse it towards extracting a meaningful pattern and detecting a gesture.

Methods for motion-based recognition of human gestures proposed in the literature [79] have often been developed to deal with sequences from a single perspective [82]. Considerably less work has been published on recognising human gestures using multiple cameras. Mono-ocular human gesture recognition systems usually require motion to be parallel to the camera plane and are very sensitive to occlusions. On the other hand, multiple viewpoints allow exploiting spatial redundancy, overcome ambiguities caused by occlusion and provide 3D position information as well.

Detection of simple features: A first approach to gesture detection is to process the input data, that is, multiple images, and extract some descriptors related with the motion of the person. Motion descriptors introduced by [82] and extended by [83] have been extensively used for motion-based gesture recognition (see an example in Figure 5.5). These descriptors generate an image/volume capturing an accumulation of the motion that happened in the last N frames and the evolution of this motion within this lapse. Other features that have been used for HCI applications are based on body silhouette analysis and crucial points extraction. For instance, in [84], this approach is used for an interface with an augmented reality game.

Articulated body models: Exploiting the underlying articulated structure of the human body allows fitting a human body model to the input data towards estimating the pose of the subject under study. The temporal evolution of this pose can be analysed towards

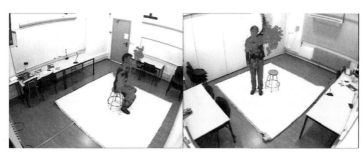

(a) Motion energy volume

(b) Motion history volume

FIGURE 5.5   Example of motion descriptors. In (a) and (b) the 2D projections of motion energy volumes (MEV) and motion history volumes (MHV) are depicted, respectively, for gestures *sitting down* and *raising hand*.

detecting gestures. An example is shown in Fig. 5.6. The pose estimation using information provided by multiple cameras has two main issues to be reviewed: the model fitting algorithm and the employed data. As it has been mentioned, a model of the human body is fit to the input data, and this model has a number of defining parameters, typically its centre, overall rotation and the values of the angles encoded at each body joint. The number of parameters for an enough detailed model may range from 10 to more than 70, and the cost function relating this model with the input data tends to have multiple minima and a high dimensionality. Unfortunately, techniques based on linear approximation and tracking algorithms, i.e., Kalman, tend to not deliver an accurate fitting. Instead, algorithms relying on Monte Carlo techniques proved to work in these circumstances with an affordable computation

FIGURE 5.6   Example of motion tracking based on articulated body models using multiple cameras in a boxing sequence.

time. There exist several possible strategies to reduce the complexity of the problem based on refinements and variations of the seminal particle filtering idea [69]. MacCormick [85] presented partitioned sampling as a highly efficient solution to this problem. Hierarchical sampling presented by Mitchelson [86] tackles the dimension problem by exploiting the human body structure and hierarchically exploring the state space. Finally, annealed Particle Filters (PF) presented by Deutscher [87] is one of the most general solutions to the problem of dimensionality. This technique uses a simulated annealing strategy to concentrate the particles around the peaks of the likelihood function by propagating particles over a set of progressively smoothed versions of the likelihood functions, thus avoiding getting trapped in local maxima.

Regarding the input data to be fed to the fitting algorithms, we may differentiate two main cases: marker-based and markerless. However, the marker-based approach is somehow intrusive thus not being appropriate to design HCI. Systems based on a markerless approach [85–87] take the multicamera video streams as the input for their tracking algorithm. In these cases, a number of instances of the human body model are generated (particles), and their fitness is measured against some features extracted from these images taking into account calibration information. For instance, the methods used in [86, 87] extract silhouettes and edges in the images and measure the particle likelihood (fitness) as a linear combination between the silhouette overlap of the body model projected onto the images and the distance between the detected edges and the projected ones from the body model. Some other approaches to this problem build up a 3D reconstruction of the scene and generate their measurements directly on this space as done in [58].

5.6 CONCLUSIONS

In this chapter, we have reviewed image and video analysis tools used for HCI. Face and hand analysis techniques that can be used in close view interfaces, such as desktop computer applications, have been described. Applications for face and hand analysis include face recognition, facial expression recognition and gaze estimation. Examples of these applications have also been described. Most of these techniques are restricted not only to close-view frontal faces but also to predetermined hand poses and gestures.

Multiple camera scenarios allow to develop HCI applications for more general situations. Head and body pose can be determined more accurately when using multiple points of view, solving occlusion or perspective problems. In this context, we have first reviewed the tools used for head orientation estimation, and how they are used for FoA estimation applications. Finally, techniques for body gesture analysis have been described. Although simple features can be used for recognising some specific actions, more complex analysis using articulated body models is needed for a complete knowledge of the human motion.

REFERENCES

1. M. Yang, D. Kriegman, N. Ahuja, Detecting faces in images: a survey, IEEE Trans. PAMI, 24 (1) (2002) 34–58.
2. E. Hjelmas, B. Low, Face detection: a survey, Comput. Vis. Image Underst. 83 (2001) 236–274.
3. H. Rowley, S. Baluja, T. Kanade, Neural network-based face detection, IEEE Trans. Pattern Anal. Mach. Intell. 20 (1) (1998) 23–38.

4. E. Osuna, R. Freund, F. Girosi, Training support vector machines: An application to face detection, in: Proceedings of Computer Vision and Pattern Recognition, Puerto Rico, (1997), pp. 130–136.

5. D. Roth, M. Yang, N. Ahuja, A SNoW-based face detector, Adv. Neural Inf. Process. Syst. 12 (NIPS12) (2000), pp. 855–861.

6. H. Schneiderman, Learning statistical structure for object detection, in: Computer Analysis of Images and Patterns (CAIP), (2003), pp. 434–441.

7. P. Viola, M. Jones, Rapid object detection using a boosted cascade of simple features, in: Proceedings of IEEE Conference on Computer Vision and Pattern Recognition, Kauai, HI, December 2001.

8. F. Fleuret, D. Geman, Coarse-to-fine face detection, Int. J. Comput. Vis. 41 (2001) 85–107.

9. D. Chai, K. Ngan, Face segmentation using skin-color map in videophone applications, IEEE Trans. Circuits Syst. Video Technol. 9 (4) (1999) 551–564.

10. C. Garcia, G. Tziritas, Face detection using quantized skin color regions merging and wavelet packet analysis, IEEE Trans. Multimed. 1 (3) (1999) 264–277.

11. V. Vilapana, F. Marques, P. Salembier, Binary partition trees for object detection, IEEE Trans. Image Process. 17 (11) (2008) 2201–2216.

12. K. Yow, R. Cipolla, Feature-based human face detection, Image Vis. Comput. 15 (9) (1997) 713–735.

13. D. Maio, D. Maltoni, Real-time face location on gray-scale static images, Pattern Recogni. 33 (9) (2000) 1525–1539.

14. R. Lienhart, J. Maydt, An extended set of haar-like features for rapid object detection, in: Proceedings of IEEE International Conference on Image Processing, (2002), pp. 900–903.

15. I. Fasel, B. Fortenberry, J. Movellan, A generative framework for real-time object detection and classification, Comput. Vis. Image Underst. 98 (2005) 182–210.

16. A. Yilmaz, O. Javed, M. Shah, Object tracking: a survey, ACM Comput. Surv. 38 (4) 2006.

17. A. Bagdanov, A.D. Bimbo, W. Nunziati, Improving evidential quality of surveillance imagery through active face tracking, in: IEEE International Conference Pattern Recognition (ICPR 2006), Hong Kong, August 2006.

18. S. Gong, S. McKenna, A. Psarrou, Dynamic Vision, From Images to Face Recognition, Imperial College Press, London, UK, 2000.

19. R.C. Verma, C. Schmid, K. Mikolajczyk, Face detection and tracking in a video by propagating detection probabilities, IEEE Trans. Pattern Anal. Mach. Intell. 25 (10) (2003) 1215–1228.

20. T. Cootes, G. Edwards, C. Taylor, Active appearance models, IEEE Trans. Pattern Anal. Mach. Intell. 23 (2001) 681–685.

21. G. Bradsky, Computer vision face tracking for use in a perceptual user interface, in: IEEE Workshop on Applications of Computer Vision, Princeton, N.J., 1998, pp. 214–219.

22. V. Vilaplana, F. Marques, Region-based mean shift tracking: application to face tracking, in: IEEE International Conference on Image Processing ICIP'08, 2008, pp. 2712–2715.

23. M. Pardàs, E. Sayrol, Motion estimation based tracking of active contours, Pattern Recognit. Lett. 22 (2001) 1447–1456.

24. D. Magee, B. Leibe, On-line face tracking using a feature-driven level set, in: British Machine Vision Conference (BMVC 2003), Norwich, UK, September 2003.

25. M. Valstar, M. Pantic, Fully automatic facial action unit detection and temporal analysis, in: Proceedings of International Conference on Computer Vision and Pattern Recognition, vol. 3, 2006, pp. 149–156.

26. M. Turkan, M. Pardas, A. Cetin, Edge projections for eye localization, Opt. Eng. 47 (2008) 1–6.

27. D. Terzopoulos, K. Waters, Analysis and synthesis of facial image sequences using physical and anatomical models, IEEE Trans. Pattern Anal. Mach. Intell. 15 (6) (1993) 569–579.

28. L. Wiskott, J. Fellous, N. Krüger, C.V. der Malsburg, Face recognition by elastic bunch graph matching, IEEE Trans. Pattern Anal. Mach. Intell. 19 (7) (1997) 775–779.

29. M.B. Stegmann, B.K. Ersbøll, R. Larsen, FAME – a flexible appearance modelling environment, IEEE Trans. Med. Imaging, 22 (10) (2003) 1319–1331.

30. M. Hamouz, J. Kittler, J. Kamarainen, P. Paalanen, H. Kalviainen, J. Matas, Feature-based affine-invariant localization of faces, IEEE Trans. Pattern Anal. Mach. Intell. 27 (9) (2005) 1490–1495.

31. D. Cristinacce, T. Cootes, I. Scott, A multi-stage approach to facial feature detection, in: 15th British Machine Vision Conference, London, England, 2004, pp. 277–286.

32. H. Tao, T. Humang, Explanation-based facial motion tracking using a piecewise bezier volume deformation model, in: Proceedings of International Conference on Computer Vision and Pattern Recognition, 1999.

33. C. Morimoto, M. Mimica, Eye gaze tracking techniques for interactive applications, Comput. Vis. Image Underst. 98 (1) (2005) 4–24.

34. Y. Ebisawa, M. Ohtani, A. Sugioka, Proposal of a zoom and focus control method using an ultrasonic distance-meter for video-based eye-gaze detection under free-head conditions, in: Proceedings of the 18th International Conference of the IEEE Engineering in Medicine and Biology Society, 1996.

35. Q. Ji, H. Wechsler, A. Duchowski, M. Flickner, Eye detection and tracking, Computer Vision and Image Understanding, 98 (2005) special issue, pp. 1–210.
36. R. Valenti, T. Gevers, Accurate eye center location and tracking using isophote curvature, in: Computer Vision and Pattern Recognition, 2008, pp. 1–8.
37. W. Zhao, R. Chellappa, P. Phillips, A. Rosenfeld, Face recognition: a literature survey, ACM Comput. Surv. 35 (4) (2003) 399–458.
38. M. Turk, A. Pentland, Eigenfaces for recognition, J. Cogn. Neurosci. 3 (1) (1991) 71–86.
39. A. Pentland, B. Moghaddam, T. Starner, View-based and modular eigenspaces for face recognition, in: Conference on Computer Vision and Pattern Recognition, Seattle, June 1994.
40. B. Moghaddam, Principal manifolds and probabilistic subspaces for visual recognition, IEEE Trans. Pattern Anal. Mach. Intell. 24 (6) (2002) 780–788.
41. P.N. Belhumeur, J.P. Hespanha, D.J. Kriegman, Eigenfaces vs. fisherfaces: recognition using class specific linear projection, IEEE Trans. Pattern Anal. Mach. Intell. 19 (7) (1997) 711–720.
42. B. Scholkopf, A. Smola, K. Muller, Nonlinear component analysis as a kernel eigenvalue problem, Neural Comput. 10 (3) (1998) 1299–1319.
43. J. Tenenbaum, V. de Silva, J. Langford, A global geometric framework for nonlinear dimensionality reduction, Science, 290 (2000) 2319–2323.
44. S. Zhou, V. Krueger, R. Chellappa, Probabilistic recognition of human faces from video, Comput. Vis. Image Underst. 91 (1–2) (2003) 214–245.
45. Y. Tian, T. Kanade, J. Cohn, Facial expression analysis, Handbook of Face Recognition, 2005, pp. 247–276.
46. R. Cowie, E. Douglas-Cowie, N. Tsapatsoulis, G. Votsis, S. Kollias, W. Fellenz, et al., Emotion recognition in human-computer interaction, IEEE Signal Process. Mag. 18 (1) (2001) 32–80.
47. B. Fasel, J. Luettin, Automatic facial expression analysis: a survey, Pattern Recogni. 36 (1) (2003) 259–275.
48. M. Pantic, L. Rothkrantz, Automatic analysis of facial expressions: The state of the art, IEEE Trans. Pattern Anal. Mach. Intell. 22 (12) (2000) 1424–1445.
49. A. Young, H. Ellis, Handbook of Research on Face Processing. Elsevier, 1989.
50. M. Bartlett, G. Littlewort, M. Frank, C. Lainscsek, I. Fasel, J. Movellan, Fully automatic facial action recognition in spontaneous behavior, in: Proceedings of the Conference on Face & Gesture Recognition, 2006, pp. 223–230.
51. N. Sebe, M. Lew, Y. Sun, I. Cohen, T. Gevers, T. Huang, Authentic facial expression analysis, Image Vis. Comput. 25 (12) (2007) 1856–1863.
52. P. Ekman, W. Friesen, J. Hager, Facial action coding system, Consulting Psychologists Press, Palo Alto, CA, 1978.

53. M. Pardàs, A. Bonafonte, Facial animation parameters extraction and expression recognition using Hidden Markov Models, Signal Process. Image Commun. 17 (9) (2002) 675–688.
54. J. Lien, T. Kanade, J. Cohn, C. Li, Detection, tracking, and classification of action units in facial expression, Rob. Auton. Syst. 31 (3) (2000) 131–146.
55. F. De la Torre, Y. Yacoob, L. Davis, A probabilistic framework for rigid and non-rigid appearance based tracking and recognition, in: Fourth IEEE International Conference on Automatic Face and Gesture Recognition, 2000. Proceedings, 2000, pp. 491–498.
56. Y. Yacoob, L. Davis, Computing spatio-temporal representations of human faces, in: 1994 IEEE Computer Society Conference on Computer Vision and Pattern Recognition, 1994. Proceedings CVPR'94, 1994, pp. 70–75.
57. T. Moeslund, E. Granum, A survey of computer vision-based human motion capture, Comput. Vis. Image Underst. (2001), 231–268.
58. C. Canton-Ferrer, J. Casas, M. Pardas, Exploiting structural hierarchy in articulated objects towards robust motion capture, Lect. Notes Comput. Sci. 5098 (2008) 82–91.
59. A. Erol, G. Bebis, M. Nicolescu, R. Boyle, X. Twombly, Vision-based hand pose estimation: a review, Comput. Vis. Image Underst. 108 (1–2) (2007) 52–73.
60. V. Pavlovic, R. Sharma, T. Huang, Visual interpretation of hand gestures for human-computer interaction: a review, IEEE Trans. Pattern Anal. Mach. Intell. 19 (7) (1997) 677–695.
61. J. Letessier, F. Bérard, Visual tracking of bare fingers for interactive surfaces, in: UIST, vol. 4, 2004, pp. 119–122.
62. X. Zhu, J. Yang, A. Waibel, Segmenting hands of arbitrary color, in: Proceedings of the Fourth IEEE International Conference on Automatic Face and Gesture Recognition 2000. IEEE Computer Society Washington, DC, USA, 2000.
63. S. Soutschek, J. Penne, J. Hornegger, J. Kornhuber, 3-D gesture-based scene navigation in medical imaging applications using Time-of-Flight cameras, in: IEEE Computer Society Conference on Computer Vision and Pattern Recognition Workshops, 2008. CVPR Workshops 2008, 2008, pp. 1–6.
64. S. Guomundsson, R. Larsen, H. Aanaes, M. Pardas, J. Casas, TOF imaging in Smart room environments towards improved people tracking, in: IEEE Computer Society Conference on Computer Vision and Pattern Recognition Workshops, 2008. CVPR Workshops 2008, 2008, pp. 1–6.
65. S. Conseil, S. Bourennane, L. Martin, Comparison of Fourier Descriptors and Hu Moments for Hand Posture, in: European Signal Processing Conference (EUSIPCO), 2007.
66. R. Hartley, A. Zisserman, Multiple View Geometry in Computer Vision. Cambridge University Press, 2004.

67. E. Murphy-Chutorian, M. Trivedi, Head pose estimation in computer vision: a survey, IEEE Trans. Pattern Anal. Mach. Intell. 31 (4) (2009) 607–626.

68. M. Chen, A. Hauptmann, Towards robust face recognition from multiple views, in: Proceedings of IEEE International Conference on Multimedia and Expo, 2004.

69. M. Arulampalam, S. Maskell, N. Gordon, T. Clapp, A tutorial on particle filters for online nonlinear/non-Gaussian Bayesian tracking, IEEE Trans. Signal Process. 50 (2) (2002) 174–188.

70. C. Canton-Ferrer, C. Segura, J. Casas, M. Pardàs, J. Hernando, Audiovisual head orientation estimation with particle filters in multisensor scenarios, EURASIP J. Adv. Signal Process. 2008, pp. 1–12.

71. CLEAR – Classification of Events, Activities and Relationships Evaluation and Workshop. <http://www.clear-evaluation.org>, 2007.

72. M. Voit, K. Nickel, R. Stiefelhagen, Head pose estimation in single and multi-view environments – results on the CLEAR'07 benchmarks, in: Proceedings CLEAR Evaluation Workshop, 2007.

73. S. Yan, Z. Zhang, Y. Fu, X. Hu, T. Jilin, T. Huang, Learning a person-independent representation for precise 3D pose estimation, in: Proceedings CLEAR Evaluation Workshop, 2007.

74. Z. Zhang, G. Potamianos, A. Senior, T. Huang, Joint face and head tracking inside multi-camera smart rooms, Signal, Image Video Process. 1 (2) (2007) 163–178.

75. C. Canton-Ferrer, C. Segura, M. Pardas, J. Casas, J. Hernando, Multimodal real-time focus of attention estimation in SmartRooms, in: IEEE Computer Society Conference on Computer Vision and Pattern Recognition Workshops, 2008. CVPR Workshops 2008, 2008, pp. 1–8.

76. R. Stiefelhagen, J. Yang, A. Waibel, Modelling focus of attention for meeting indexing based on multiple cues, IEEE Trans. Neural Netw. 13 (4) (2002) 928–938.

77. M. Argyle, M. Cook, Gaze and Mutual Gaze. Cambridge University Press, 1976.

78. R. Stiefelhagen, J. Yang, A. Waibel, Modelling people's focus of attention, in: Proceedings IEEE International Workshop on Modelling People, 1999.

79. J.K. Aggarwal, Q. Cai, Human motion analysis: a review, in: Proceedings of the IEEE Nonrigid and Articulated Motion Workshop, 1997, pp. 90–102.

80. M. Black, F. Brard, A. Jepson, W. Newman, W. Saund, G. Socher, M. Taylor, The Digital Office: Overview, in: Proceedings of Spring Symposium on Intelligent Environments, 1998, pp. 98–102.

81. CHIL – Computers in the Human Interaction Loop. <http://chil.server.de>, 2004–2007.

82. A. Bobick, J. Davis, The recognition of human movement using temporal templates, IEEE Trans. Pattern Anal. Mach. Intell. 23 (3) (2001) 257–267.

83. C. Canton-Ferrer, J. Casas, M. Pardàs, Human model and motion based 3D action recognition in multiple view scenarios, in: Proceedings of European Signal Processing Conference, 2006.

84. P. Correa, J. Czyz, F. Marques, T. Umeda, X. Marichal, B. Macq, Bayesian approach for morphology-based 2-D human motion capture, IEEE Trans. Multimed. 9 (4) (2007) 754–765.

85. J. MacCormick, M. Isard, Partitioned sampling, articulated objects, and interface-quality hand tracking, in: Proceedings of European Conference on Computer Vision, 2000, pp. 3–19.

86. J. Mitchelson, A. Hilton, Simultaneous pose estimation of multiple people using multiple-view cues with hierarchical sampling, in: Proceedings of British Machine Vision Conference, 2003.

87. J. Deutscher, I. Reid, Articulated body motion capture by stochastic search, Int. J. Comput. Vis. 61 (2) (2005) 185–205.

Processing of Handwriting and Sketching Dynamics

Claus Vielhauer

Brandenburg University of Applied Science, Otto-von-Guericke University, Magdeburg, Germany

6.1 INTRODUCTION

The intention of this chapter is twofold. In the first parts (Sections 6.2–6.5), an overview to the modality of handwriting and sketching and the resulting fundamentals in signal processing in the context of human–computer interaction (HCI) is given. The second parts 6.6 and 6.7 review selected technical methods for two exemplary applications

Multimodal Signal Processing, ISBN: 9780123748256

for pen-based HCI: sketch recognition for user interface design and similarity search in electronic ink documents.

For the first part, Section 6.2 briefly reflects the evolution of writing as a human capability, its demise over the industrial age and its renaissance during the computer age by first ideas for pen-based user input devices such as the light pen. Section 6.3 then presents some main principles in the digital acquisition of handwriting dynamics and reviews some relevant concepts for digitiser tablet sensor technology arising from them. Principles in signal acquisition by digitiser tablets, such as the physical phenomena and the technical concepts for acquiring pen movement signals are discussed in Section 6.4, with a focus on optical, electromagnetic and resistive/capacitive film-based tablets. Digital representations for typical pen movement signals, such as horizontal and vertical pen movement, pen-tip pressure signals, are discussed in Section 6.4. Section 6.5 identifies some relevant problems that can be solved by means of handwriting modality in HCI, including different applications for text, shape and gesture recognition, as well as biometric user authentication. Further, Section 6.5 introduces a generalised and application independent model for the understanding of feature-based recognition.

The first section of the more technically detailed second part of this chapter discusses, in Section 6.6, different methods as how the recognition of hand-drawn sketches from digitiser tablets can be achieved to support users in user interface (UI) design. In this section, emphasis is given to technical outlines and comparison of three different algorithms: the *CALI library, Rubine's algorithm* and the trainable recogniser in *SketchiXML*.

A selection of approaches to search in repositories of so-called Digital Ink, i.e., digital notes taken from digitiser tablets and stored in digital databases, are introduced in Section 6.7. Here, examples for digital note taking applications available today for different operating systems are presented, and the necessity of fuzzy search in data sets generated by the use of such systems is motivated. Because of the observation that search in textual space, preceded by handwriting recognition, is not feasible in many applications, this section further introduces an alternative concept, which is based on the alphanumeric coding of four different feature types: *Freeman codes*, *direction*, *curvature* and *slant angle* of the writing trace. All four different types are illustrated and experimental results regarding the recognition accuracy in terms of precision and recall rates conclude this section.

The last part of this chapter provides a summary and an outline of current research directions in Section 6.8.

6.2   HISTORY OF HANDWRITING MODALITY AND THE ACQUISITION OF ONLINE HANDWRITING SIGNALS

Doubtlessly, the development of handwriting capability is one of the greatest achievements of the evolution of mankind. From a historical perspective and to our knowledge today, human beings started about 3 millennia ago to use cuneiform script, as the first form of abstract representation of knowledge. This form of script has developed from figurative pictograms towards a more complex system of modelling phonetic components of speech by symbol sets, i.e., alphabets, and therefore can be seen as a great innovation for persistently storing and communicating knowledge [1]. Thus, the progress towards handwriting has established a new means in human-to-human communications and a kind of offline interaction, without the necessity of communication partners to be present at the same place and time. However, handwriting had originally been limited by the number of physical copies that could be produced during a single writing process and consequently, through the centuries, techniques for printing, i.e., generating multiple copies of documents from one source, have been developed. Early methods based on wood blocks have been reported from ancient Asia (as from app. 751 AD) where reliefs on closed areas of wooden surfaces have been used as a negative to print ink on paper or textiles. These initial techniques have been further refined towards moveable type printing. Examples for wooden types have been observed in Korea from the 11*th* century AD, with first mentions of metal types in 1234. This technique, allowing more flexibility due to the possibility of casting together the printing text from sets of letter punches, has been introduced in Europe much later, in the mid 15*th* century, by J. Gutenberg [2]. With the more recent introduction of typewriters (19*th* century) and the breakthrough of the first generation of modern personal computers and digital printing technology in the last century, it appeared almost inevitable, that the relevance of handwriting in modern societies will dramatically shrink.

However, as part of the development of modern computer equipment, researcher and scientists considered the adaptation of the concept of using pens for drawings and handwriting for HCI already at a relatively early stage. One of the first acquisition methods to enable pen-based user input to computers has been the so-called light pen technique [3]. Here, an electronic pen device is equipped with a light-sensitive sensor at the pen tip, to be used to write on, or close to, the surface of a monitor based on cathode ray technique. Synchronised with the line scanning sequence of the cathode ray, this concept allows for relatively straightforward computation of the pen tip position by measurement of the exact time of the cathode ray passing the light sensor, as illustrated in Figure 6.1.

One of the first applications of HCI based on such light pen technologies has been the Sketchpad project in the 1960's [3], whereby Sutherland describes a system and interaction concept for generating and editing drawings based on geometric structures on a computer system.

Looking at today's zoo of computer devices, which has emerged from those days, we may observe a renaissance trend. Many of the modern computers, and among these particularly those designed for mobile use, such as personal digital assistants (PDA), smart mobile phones, portable notebook computers (e.g., Tablet PCs), as well as computer displays are equipped with different techniques for pen-based user interaction. Also, a great variety of software applications have emerged, ranging from pen-based drawing, handwriting recog-

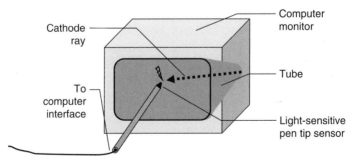

FIGURE 6.1   Simplified illustration of the light pen concept, from [4].

nition, sketching, gesturing all the way to biometric user recognition by signature analysis. This chapter provides an overview to technical concepts of the most common sensors today, reviews principles for the analysis of signals derived from writing and drawing dynamics and discusses different recognition objectives. Finally, we briefly review algorithms for two exemplary applications: pen-based sketches for interactive interface design [5] and similarity search in databases of handwriting digital documents [6].

6.3   BASICS IN ACQUISITION, EXAMPLES FOR SENSORS

The variety of sensors for sampling pen-based dynamics, also referred to as digitiser tablets, which is available today is based on different physical phenomena. This section provides a short review of acquisition technologies and sensors, based on optical, electromagnetic and resistive/capacitive film methods.

Besides the aforementioned light pen technology, there exists at least one more concept for optically tracing the pen movement, which is based on a defined pattern structure, which encode the position by almost invisible patterns on pre-printed paper. In this technology, pens are equipped with tiny digital cameras, integrated in the pen tip, which continuously perform video recordings of the area below the pen tip. A pattern recognition algorithm, executed by a computer device embedded in the pen, determines the pen position based on co-ordinates encoded by the patterns printed on the pen surface. Sensors of this type have been developed, commercialised and licensed to other organisations for example by a company named Anoto AB (www.anoto.com).

Besides more exotic methods based on ultrasonic triangulation, acceleration and force sensors attached to the writing surface, today the majority of digitiser tablets are either electromagnetically operated or are based on resistive/capacitive film technology. The first category is typically based on inductive coupling between an inductor in the pen tip and a matrix of circuit tracks embedded in the writing surface (e.g., the tablet PC display), powered by AC voltage at different frequencies. By frequency analysis of the voltage induced at the pen tip, the pen position can be determined. Tablets based on resistive

field technology take advantage of measurement of potential differences between a conductible film surface and a conductible pen tip. Here, the surface is actuated by a potential difference, e.g., from left to right for determination of the horizontal position and for a linear voltage curve along the plane, the determination of the first dimension of the pen position is straightforward. By temporally altering between horizontal and vertical powering, the second dimension can be obtained in the same manner. Similar to resistive field technology, but relying on capacitive effects between two conductive film layers, another category of sensors is composed of a grid of pressure-sensitive capacitors, made of equally distanced dot spaces between two layers of film.

In addition to the pen position, most of the modern digitiser tablets possess sensors to determine, whether or not the pen tip is pressed on the surface. Here, two main concepts can be found: in the more simple case, a pushbutton-type switch is mechanically attached to the pen tip, simply providing pen-up/pen-down information to the application. In the more sophisticated case, tablets are able to quantise the degree of pressure exposed to the pen tip by means of pressure sensors. These can be either integrated in the pen tip (in case of electromagnetic systems) or they are provided by voltage measurement in case of capacitive film technology.

Note that the sensoring schemes presented so far cover exclusively the so-called *online* domain, in contrast to *off-line* handwriting acquisition, where handwritten documents are optically scanned after completion of the writing process. Online approaches are considered to be more appropriate for recognition than off-line ones, because of the availability of process-related data, such as stroke direction and ordering, as well as timing information [7, 8]. Off-line methods, however, are limited to image processing, where only the result of the writing process can be processed.

6.4 ANALYSIS OF ONLINE HANDWRITING AND SKETCHING SIGNALS

The technological concepts summarised earlier should reflect the most common ones today, although they are non-exhaustive. However, independently of the underlying sensor technology, all digitiser tablets

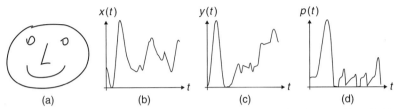

FIGURE 6.2   Sketching of a symbol (a) and the resulting signals $x(t)$, $y(t)$ and $p(t)$ from the digitiser tablet in parts (b), (c) and (d), respectively.

address one central goal: tracing the pen movement during writing, drawing, sketching or signature processes over time. Consequently, all of these devices provide at least two categories of position signals: the horizontal and the vertical position of the pen tip over time. In accordance to the image processing notation, these signals are also usually referred to as $x(t)$ (horizontal) and $y(t)$ (vertical) writing position signal [9]. Also the great majority of devices measure pen pressure over time (often denoted as $p(t)$), either as a binary value for pen-up/pen-down or as a quantised pressure value. Figure 6.2 illustrates an example of $x(t)$, $y(t)$ and $p(t)$ signals recorded during a drawing process for the symbol shown on the left.

Depending on the tablet capabilities, additional physical measurements may also be recorded by the sensors. These measurements include velocity and acceleration, as well as pen angles above the surface, etc. For the further discussions in this chapter, we limit our views to signal processing based on two dimensions ($x(t)$ and $y(t)$), because these kind of signals are common to practically all tablets and our presentations are intended to provide a general introduction to approaches to processing of handwriting and sketching dynamics. These signals can be represented as sequences of a number n of triplets (x_i, y_i, t_i), whereby t_i denotes the timestamp of measurement of the two other values and i represents the index of a triplet within the sequence of $[1, \ldots, n]$ of all n sampling points acquired from the writing trace.

6.5   OVERVIEW OF RECOGNITION GOALS IN HCI

The signal data from pen movements as introduced earlier can be the origin for multiple alternative recognition goals in HCI applications.

In general, these recognition goals can be characterised by the type of information, which users intend to document by their writing or drawing action. A straightforward distinction here is *textual/non-textual content*, whereby textual data can be seen as sets of characters, numbers and punctuation marks [10].

On the basis of the writing method, textual input can be further subdivided into two kinds of system. Character-wise input include systems, where users have to input the text character by character, in pre-defined fields on the tablet whereby either a pre-defined or user-adaptive set of gestures is related to each alphabet symbol. The other category of textual input is commonly known as cursive script recognition, whereby entire sequences of connected handwriting (e.g., words) are recognised rather than individual characters.

Examples for this first category are EdgeWrite [11] and Graffiti™, both of which are based on unistroke symbols, i.e., the pen performs the writing sequence for one character in a single stroke, i.e., sequences of continuous pen movement, with no pen lift interrupts. Note the company who originally introduced it, Palm™ Inc., no longer distributes the later system after a patent case. For the second category, a great variety of methods have been suggested in the past, overviews of research activity are given in [8] and [12].

With respect to non-textual input, three main sub-categories of recognition applications can be formed: gesture recognition, symbols and sketches as well as biometric user recognition. Gestures are pen movements, which are interpreted as commands to computer applications or operating systems (OS). Examples for this category are interactive pen gestures for encircling, scratching or moving of graphical objects in pen-enabled OS. A specific characteristic of gesture recognition is that the pen movements do not leave any digital ink traces after the recognition, which is the case for *symbol and sketch recognition*. Here, the goal is to recognise symbols from a given set to form a pen-drawn sketch representation of higher complexity. A wide variety of applications exist for this domain, and for the sake of briefness, we limit to the following mentions: the detection of linguistic symbols [13, 14] for proofreading of documents, the interpretation of hand-drawn graphical widgets and automated generation of program code from those [5] and also recognising and simulating electrical circuits from hand-drawn symbols [15].

Finally, *biometric user authentication* can be achieved by analysing the signals of an actually performed handwriting action (in this case usually the signature) and comparing properties from these against user specific reference data. If the comparison yields sufficient similarity, it can be assumed that the identity of the actual writer is identical to the person who originally provided the reference data; the user can be authenticated. Again for this application, a great variety of methods have been suggested in the domain of so-called online signature verification, an overview can be found for example in [9].

Regardless of the recognition goals of online handwriting, the majority of recognition algorithms follow the concept of pattern recognition. The basic idea behind this concept is illustrated in a model in Figure 6.3: characteristic features are extracted by signal processing on the raw sensor data, after an optional pre-processing, e.g., for noise suppression. Once signals have been transformed in feature space, some comparison method assigns the input signals to one of the class templates in the reference data set, provided a sufficient degree of conformance exists. Note that the reference samples can either be static, i.e., predefined by the system, or user-specific. The later cases are often referred to as user-trainable recognising techniques.

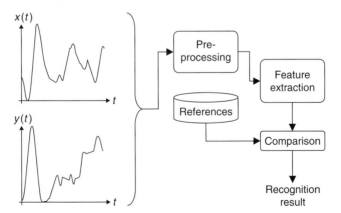

FIGURE 6.3   Generalised model for recognising handwriting by feature-oriented signal processing a pattern recognition scheme.

For use in HCI and each of the other aforementioned recognition goals, an extraordinarily wide variety of methods and combinations for pre-processing, feature extraction and comparison have been suggested by researchers over the past decades. Rather than elaborating on a survey of these approaches, the aim of this chapter is to give the reader an idea about how handwriting-based recognition in HCI can be technically achieved by means of examples. We therefore introduce two exemplary methods in the remaining sections of this chapter, having two different recognition goals. The first one addresses the domain of the aforementioned *sketch recognition* and summarises selected approaches to a user-adaptive extension for recognising widgets for a pen-input-based interface design tool.

With the growing number of digital documents, including documents produced in digital ink, the problem of searching and finding data in large databases will become increasingly evident. Therefore, the second example in this chapter reviews a method for searching in archives of digital handwriting data, without the necessity to perform a handwriting to text recognition and with the capability to search symbols as well as connected text sequences. The later example implements a kind of query-by-example interface for retrieval of handwritten content.

6.6   SKETCH RECOGNITION FOR USER INTERFACE DESIGN

An important aspect in computer science is the design of appropriate UI for applications and to support this process by technical means such as paper sketching, prototypes, mock-ups and diagrams. This section reflects a selection of three methods for this purpose (CALI library, Rubine's algorithm and SketchiXML) and discusses their properties.

For many researchers, the concept of hand sketches on paper appears to be the most intuitive way for designers to draft a first layout of a UI, see for example [16], mainly because of its intuitional and largely unconstrained character. These conclusions have led to the idea of computer-based sketching UI tools, which are based on pen input on digitiser tablets and subsequent signal processing. Two main categories can be identified in this domain: approaches that support the sketching only, without interpreting it and secondly, methods that also try to interpret the hand-drawn input to recognise its meaning. For the second category, a wide range of approaches to recognise geometric forms, widgets and other graphical objects have been suggested, a structured overview of this variety is presented in [5]. In this chapter, we focus on one specific subcategory within these recognisers, that is systems that process sketched data on tablet directly into a coded digital representation. Among others, SILK [16], JavaSketchIt [17], FreeForm [18, 19] and SketchiXML [20] are systems that pursue this goal, whereby the specific output to be produced varies from code for operating systems, source code for programming languages all the way to XML descriptions for UIs such as UsiXML (User Interface eXtensible Markup Language – http://www.usixml.org).

With respect to the underlying signal processing and recognition algorithms, two main concepts can be found: the *CALI library* [21] for shape recognition and *Rubine's algorithm* [22] for gesture recognition.

The first is based on an algorithm to recognise seven different geometrical shapes (shown in Figure 6.4) and additionally five command gestures (delete, wavyline, copy, move and undo) based on building a total of 14 features from the pen input signals and deriving fuzzy sets from those. Representative fuzzy sets have been determined experimentally by reference subjects and based on these tests, a rule-based decision has been coded into the library. The library is freely available in source code from the author's institution website (http://vimmi.inesc-id.pt/cali/).

Another widely recognised algorithm was published by Rubine in [22], which addresses recognition of pen-based gestures by training a model consisting of trained examples of 13-dimensional feature

FIGURE 6.4   Seven geometrical shape examples recognisable by the CALI library (from left to right): line, arrow, triangle, rectangle, circle, ellipse and diamond.

vectors $(f_1,...,f_{13})$ composed of statistical features above single strokes. The latter are recorded as sequences of p triplets of pen positions $(x_i, y_i, t_i), i = [0, ..., P-1]$, with t_i denoting a time stamp. In addition, the bounding box dimensions $(x_{max} - x_{min})$ and $(y_{max} - y_{min})$ are determined as per Figure 6.5 and, from these quantities, 13 geometric and temporal features are extracted. Examples for these features are length and bounding box diagonal (f_3) and their angle with respect to the horizontal (f_4) as illustrated in Figure 6.5.

In a training phase, the system acquires feature vector samples for each of the gestures (or classes) to be recognised, represented within the system by averaged features, along with entropy-based weight factors for each of the feature vector components. During the recognition of strokes, a weighted distance is determined between the actually computed feature vector and the stored references and based on this cumulative distance, a minimum distance classification assigns the actual stroke to the closest class. Rubine's algorithm today is still very commonly used, for example by the FreeForm system [18], which generates Basic program code sequences from hand-drawn sketches.

To characterise the two approaches in summary, it can be said that the CALI method is static in a sense that it provides predefined rules with no necessity to perform user-specific training. The latter is the case for Rubine's algorithm, with an additional constraint that it is

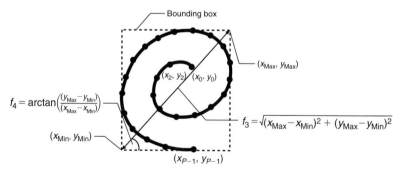

FIGURE 6.5 Example for calculation of feature f_3 and f_4 by Rubine's algorithm: (x_0, y_0) and (x_{P-1}, y_{P-1}) being the first and last pen positions in the stroke sequence, respectively. Time stamps t_i, $i = (0, ..., P-1)$ are not included in the figure.

based on single stroke sketches. Both algorithms have in common that shape recognition is invariant to the direction of the pen movement, i.e., it does not matter if for example a circle was sketched clockwise or counterclockwise.

An alternative approach for a multiple stroke trainable recogniser has been suggested in [5]. It has been successfully implemented as a secondary recogniser in the SketchiXML, in extension to the CALI system [21]. This approach is based on the decomposition of tablet signals into sequences of directional codes along the writing trace, an idea originally introduced as *Freeman chain code* [23] for compressed representations of graphical contours. In summary, the idea is based on three steps:

- Re-quantisation of the pen position triplets to the intersection points (i.e., vertices) of a square grid of fixed size, maintaining their sequential order. This process is denoted as square-grid quantisation.
- Following the stroke sequences along the grid vertices, the angle direction between each subsequent pair of pen positions is determined.
- For each of the angles computed along the stroke trace, perform a quantisation to eight possible discrete directions between two neighbour grid vertices (i.e., 0°, 45°, 90°, 135°, 180°, 225°, 270° in counter-clockwise orientation from the horizontal axis) and coding of these directions in sequences of symbols assigned to each of them.

Consequently, Freeman coding can transform any stroke sequence drawn on a digitiser tablet into a sequence of predefined symbols, a string. For the sake of simplicity, we will refer to this coding as Freeman code for the remaining sections of this chapter.

The following illustration (Figure 6.6) gives an example of this process for a simple sketch and its resulting string sequence coding, based on the inter-grid directions in sequence from (x_0, y_0) to (x_{P-1}, y_{P-1}) and a numerical alphabet of $[0, \ldots, 7]$ for each of the eight directions, as shown on page 132. Note that the hollow dots represent the sampling point after re-quantisation, whereas filled dots are the original points.

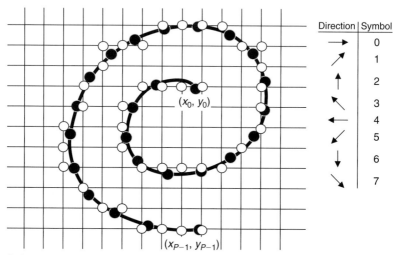

Direction	Symbol
→	0
↗	1
↑	2
↖	3
←	4
↙	5
↓	6
↘	7

String sequence code: "4445667000011222334445446565667707000"

FIGURE 6.6   Example for directional string sequence coding.

To represent multiple strokes by directional code strings, a straightforward method is to concatenate the strings of each single stroke sequence, delimited by a special symbol to characterise the gap.

By representing sketches as strings of directional codes, it is possible to use methods for string similarity measurement for comparison and recognition of sketches. A common way to compare two strings s_1 and s_2 is the string edit distance, which measures the effort to transform s_1 into s_2 by basic character-wise delete, insert and replace operations. The minimum number of such operations required for any transformation from s_1 into s_2 is also denoted as *Levenshtein distance* [24] between s_1 and s_2. Obviously, with decreasing values of Levensthein distance, the similarity of the two strings increases (two identical strings have a distance equal to zero) and increasing values denote greater dissimilarity. Many effective algorithms to determine the Levensthein distance have been suggested and it has been shown that the complexity can be reduced to a constant factor times the order of $O(\text{Max}(|s_1|, |s_2|))$, with $|s_1|, |s_2|$ denoting the string lengths of s_1 and s_2, respectively [5].

Given the string edit distance function, the design of a recogniser system can be achieved by constructing a repository of sample strings, or references, for each class of sketches during a user-specific training phase. During recognition, the signals from the digitiser tablet are again transformed into strings of directional codes and then compared to all references. The unknown sketch is then assigned to its closest match by a nearest neighbour classificator. To avoid false-positives, a minimum threshold can be introduced for class assignment.

The SketchiXML project has integrated two categories of sketch recognisers. A predefined set of widgets can be recognised from combinations of basic shapes drawn by the pen input using the CALI library method, whereas the system can seamlessly integrate enhanced recognition of widgets, shapes and commands based on a user-trainable recogniser, implemented according to the method described earlier. To date, the system is freely available from the website of the USIXML project (www.usixml.org).

6.7   SIMILARITY SEARCH IN DIGITAL INK

In the following sections, we reflect techniques for search of handwriting or symbolic sequences in electronic documents, which have been produced by users on digitiser tablets and which are stored persistently in repositories, maintaining their temporal information (i.e., online handwriting data). Methods based on textual handwriting recognition are briefly reviewed in this section, as well as techniques comparing shape contour features. Because these two concepts either rely on textual content or require well-defined word or symbol segmentation, a third technique based on string coding of different directional features is introduced, which may overcome these shortcomings. Experimental results approach based on four different feature set configurations are provided for the latter, along with a discussion of their impact to recognition accuracy.

With the increasing propagation of Tablet PCs, an increasing number of note taking applications today allow for the use of such devices in the sense of digital paper and ink, i.e., whatever freehand information a user could write or draw on real paper, is recorded by digitiser tablets and stored by the software system. The resulting type of data is also commonly known as *Digital Ink*, and there are several commercial and open-source applications available today, for example *Microsoft Windows Journal*^{TM}, *Gournal* and *Xournal* for Linux or *NoteLab* and *Jarnal* as platform independent Java implementations (http://tuxmobil.org/tablet_unix.html).

Along with the continuous use of such digital note taking applications and the digital ink documents produced this way, comes an aggregation of digital data and consequently archives of ever growing size. One of the challenges in context of HCI is to develop efficient search and retrieve techniques for such archives, similar to the powerful tools, which are available today by Web search engines. However, in comparison to the later systems, which are usually based on fuzzy search schemes applied to textual content in HTML documents, the search problem for digital ink appears differently. The problem here can be stated as a fuzzy search problem in large sets of signals, coupled with individual writing/sketching characteristics in digital representations (recall that Digital Ink Documents can be considered as a collection of triplets of pen position sampling points as introduced in Section 6.4), based on some sample information input by users. Thus, similarity search in digital ink constitutes a query-by-example problem in HCI applications, to be solved by signal processing.

To solve this problem, a number of methods have been suggested in recent research and probably the most straightforward approach is to firstly convert Digital Ink to text by means of handwriting recognition, both for the search input and the repository data and then perform search by well-studied methods of text similarity. However, because handwritten documents can be composed of handwritten content, symbols or combinations of both, such approaches, as suggested for example in [25] and [26], may not be sufficient for digital ink documents. Alternative methods suggested in the literature are based on shape comparison between query words and words in the document repository [26–28]. These approaches rely on correct word segmentation, i.e., the decomposition of entire digital ink documents into sequences of connected characters forming words, prior to the search process.

The third category of methods, which is briefly introduced here, overcomes the limitations of related work with no need to perform handwriting recognition or word/symbol segmentation [6]. The application scenario is a 'query by example' search engine, where users can draw or write a search example on the digitiser tablet, representing the search data. This search data is then compared to entire Digital Ink documents of a user-defined search set, and all parts of all documents with sufficient similarity are considered as matches. This concept allows users, for example, to search all documents in a repository containing symbols drawn in a specific manner. It is, however, intrinsically limited to user-specific queries and assumes that users write or draw the same words or symbols in a consistent order. Thus, this method addresses user-dependent search problems only and does not consider search problems across documents from different users.

Like introduced in the previous section in our reflections on trainable recognition of sketches for UI design, the basic idea is again to decompose the writing/drawing pen triplet sequences into strings of symbols for characteristic features. In [6], four different types of features have been suggested for the similarity search problem and experimentally studied with respect to their retrieval accuracy. These four alternative feature types are summarised in the following, followed by a brief discussion of their appropriateness with respect to recognition accuracy, based on the experimental results from [6].

- **Freeman Codes**: the first category of features suggested in [6] is identical to the Freeman chain coding principle as discussed in the previous section of this chapter and illustrated in Figure 6.6.
- **Direction-based Codes**: A possibility to overcome the spatial jagging due to the re-quantisation in the Freeman coding step is to measure the angle between two subsequent pen triplets, without prior spatial re-quantisation. The second feature category in [6] suggests this approach, whereby trace velocity information is neglected by a spatially equidistant resampling (i.e., regularly spaced in arc length) of the original handwriting, using cubic spline interpolation. This way, the sampling points basically remain on the original writing trace, whereas their spatial distance is normalised with respect to the path or arc length. Another advantage of this feature category is the possibility to freely configure the quantisation degree of the resulting angular directions.

- **Curvature-based Codes** are features that represent the change of curvature along the writing or drawing trace. Obviously, from a mathematical perspective, curvature is the first derivative of the two-dimensional system of equations composted by the $x(t)$ and $y(t)$ writing signals. Numerically speaking, the most straightforward way to calculate this is to again perform the same steps as for the direction-based coding and to consider the angular difference between two subsequent pairs of triplets along the writing trace as the curvature information (in this case again quantised to 16 levels).

- **Slant-based Codes** form the fourth feature category suggested in our referenced work. A slant is commonly defined as connection between a vertical minimum and the consecutive vertical maximum or a vertical maximum and the consecutive vertical minimum in the writing trace, respectively. Each slant can be measured with respect to its angular orientation and quantised to a specific number of levels.

The four different methods for feature extraction are shown in Figure 6.7, where an exemplary original sample, having non-equal arc length is shown on the left and the four different features in illustrations (b) through (d).

The actual comparison between the query samples and the documents in the repository is again performed by transforming the handwriting signals into feature space, representing data by feature

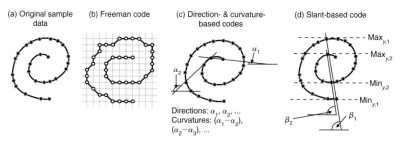

(a) Original sample data   (b) Freeman code   (c) Direction- & curvature-based codes   (d) Slant-based code

Directions: $\alpha_1, \alpha_2, \ldots$
Curvatures: $(\alpha_1 - \alpha_2)$, $(\alpha_2 - \alpha_3), \ldots$

FIGURE 6.7   Illustration of feature code calculations for the Freeman code in (b), direction- and curvature-based codes in (c) and slant-based codes in (d). α denote directional angles, β denote slant angles whereas $\mathrm{Min}_{y,1}, \mathrm{Min}_{y,2}, \mathrm{Max}_{y,1}, \mathrm{Max}_{y,1}$ denote the first and second local minima and maxima in the writing trace sequence, respectively.

strings and using string edit distance for comparison, as described in Section 6.6.

For experimental comparison of the four different feature sets, the common measures to determine accuracy in retrieval systems, precision and recall rates (PR and RR, respectively) have been experimentally calculated for a test database. PR and RR are determined as follows: given a known set of ground truth, i.e., groups of symbols or words from the same subject and having the same meaning, the sets are decomposed into query words and symbols and their correspondences in the search space. Then for each of the query sequences, searching is performed on all documents originating from the same subject and above the entire test set, the number of correct matches, mismatches and missed instances are determined. On the basis of this totalling, PR and RR can be calculated according to the following equation [29], where # denotes the number of occurrences within a test:

$$PR = \frac{\#matches}{\#matches + \#mismatches} \qquad RR = \frac{\#matches}{\#matches + \#missed\,instances}$$

$$(6.1)$$

Consequently, PR and RR values equal to 1 would imply perfect retrieval accuracy with neither mismatches nor missed sequences, where as the error tendency increases, the more RR and PR converge towards 0. Practically for the HCI application of search in Digital Ink, as discussed in this section, the PR gives an indication to which degree false matches (i.e., the search result set includes incorrect matches) in a particular system are not to be expected, i.e., how precise the system is. RR indicates the recall degree, i.e., the prevention of nonmatches.

The summary of the experimental comparison of the four different feature types based on this database is shown in Table 6.1. Note that in addition to PR and RR, the table includes the average execution time for each document in each of the 271 search cycles. Further, the table only reflects the results for the best parameterisation of features within its category. A detailed presentation of the experimental results, including a graphic of the receiver operating characteristic (ROC) of PR and RR as function of the evaluation parameters, is provided in the original publication [6].

TABLE 6.1 Experimental results for the four different feature types

	PR [%]	RR [%]	Execution time [ms]
Freeman codes	81.5	81.5	1555
Direction-based codes	69.2	72.9	1933
Curvature-based codes	32.3	32.4	176
Slant-based codes	33.1	29.3	14

It can be observed that the best accuracy regarding the search results could be achieved by Freeman Codes with $PR = RR = 81.5\%$. The curvature- and slant-based approaches, however, have shown very poor results having PR and RR in the range of 30%. Apparently these kinds of features do not preserve sufficient information in feature space to enable reasonable retrieval accuracy, at least in the matching the scheme proposed. Also, it becomes obvious that in this application scenario, recognition accuracy is gained clearly at cost of computational effort. The most accurate algorithm (Freeman) took about 100 times longer on average than the slant-based code. Compared to other related work, it appears that the best results from this test are in the same range, with a slightly better accuracy of 93.2% PR and 90% RR, reported in [27]. However, because results have been determined on different data sets and using different protocols, this comparison can only be indicative.

6.8 SUMMARY AND PERSPECTIVES FOR HANDWRITING AND SKETCHING IN HCI

In this chapter, we have presented an introduction into signal processing of handwriting and sketching in HCI by means of pen digitisers. Historical developments have been outlined in brief, along with a general review of acquisition technology and some fundamentals on signals derived from online handwriting.

On the basis of two sample applications that are of particular interest for the HCI applications, sketch recognition for user interface design and search in digital ink, we have elaborated on recent technical methods, how to process handwriting signals in application context. Also, we have presented some experimental figures for the later case, to convey some insights regarding the recognition accuracy for a modality, which is naturally bound to rather great variability.

At least three main perspectives for the handwriting modality can be seen today. The improvement of recognition accuracy remains an important aspect, which could be achieved by information fusion whereby, similarly to the already quite successful domain of biometric fusion, results of multiple subsystems – or agents – are logically combined toward improved recognition results. Further challenges are arising from modern multi-touch, multi-user tablets, allowing for new applications like collaborative sketching environments and new challenges such as user disambiguation or searches in document based on cross-subject (i.e., separate documents generated by different users) or multiple-subject (i.e., documents each of which are generated by multiple users) scope. Finally, for handwriting as a modality to identify persons, biometric cryptosystems are a current focus of research, which aim at seamlessly integrating biometric information in cryptographic protocols to increase security in terms of user authenticity.

REFERENCES

1. S.N. Kramer, The Sumerians: Their History, Culture, and Character: Their History, Culture and Character, Univ of Chicago Press, 1963.
2. T. Christensen, Gutenberg and the koreans: Did east asian printing traditions influence the european renaissance? in: Arts of Asia Magazine, Arts of Asia Publications Limited, Hong Kong, http://www.rightreading.com/printing/gutenberg.asia/gutenberg-asia-6-china-blockbook.htm.
3. I.E. Sutherland, Sketch pad a man-machine graphical communication system, in: DAC '64: Proceedings of the SHARE design automation workshop, ACM, New York, NY, USA, 1964, pp. 6.329–6.346.
4. J.E. Ward, Memorandum 5001-30, focussing light pen, Technical report, Massachusetts Institute of Technology, Cambridge 39, Massachusetts, 1960.
5. A. Coyette, S. Schimke, J. Vanderdonckt, C. Vielhauer, Trainable sketch recognizer for graphical user interface design, in: Human-Computer Interaction INTERACT 2007, 2007, pp. 124–135, Springer.

6. S. Schimke, C. Vielhauer, Similarity searching for on-line handwritten documents, J. Multimodal User Interfaces 1 (2) (2007) 49–54.

7. R. Plamondon, S.N. Srihari, On-line and off-line handwriting recognition: A comprehensive survey, IEEE Trans. Pattern Anal. Mach. Intell. 22 (1) (2000) 63–84.

8. C. Tappert, C. Suen, T. Wakahara, The state of the art in online handwriting recognition, IEEE Trans. Pattern Ana. Mach. Intell. 12 (8) (1990) 787–808.

9. C. Vielhauer, Biometric User Authentication for IT Security: From Fundamentals to Handwriting (Advances in Information Security). Springer-Verlag New York, Inc., Secaucus, NJ, USA, 2005.

10. L. Schomaker, S. Muench, K. Hartung, A taxonomy of multimodal interaction in the human information processing system: report of the esprit project 8579 Miami, Technical Report <http://hwr.nici.kun.nl/%7Emiami/>, Esprit/BRA, NICI, Nijmegen, 1995.

11. J.O. Wobbrock, B.A. Myers, J.A. Kembel, Edgewrite: a stylus-based text entry method designed for high accuracy and stability of motion, in: UIST '03: Proceedings of the 16th annual ACM symposium on User interface software and technology, ACM, New York, NY, USA, 2003, pp. 61–70.

12. G. Dimauro, S. Impedovo, G. Pirlo, A. Salzo, Handwriting recognition: State of the art and future trends, in: BSDIA '97: Proceedings of the First Brazilian Symposium on Advances in Document Image Analysis, Springer-Verlag, London, UK, 1997, pp. 1–18.

13. M.L. Coleman, Text editing on a graphic display device using hand-drawn proofreader's symbols, in: Pertinent Concepts in Computer Graphics, Proceedings of the Second Univeristy of Illinios Conference on Computer Graphics, 1969, pp. 282–290.

14. J. Andre, H. Richy, Paper-less editing and proofreading of electronic documents, in: Proceedings of EuroTex'99, 1999.

15. R. Davis, Magic paper: sketch-understanding research, Comput. 40 (9) (2007) 34–41.

16. J.A. Landay, B.A. Myers, Sketching interfaces: toward more human interface design, Comput. 34 (3) (2001) 56–64.

17. A. Caetano, N. Goulart, M. Fonseca, J. Jorge, Javasketchit: Issues in sketching the look of user interfaces, in: AAAI Spring Symposium on Sketch Understanding, AAAI Press, Menlo Park, 2002, pp. 9–14.

18. B. Plimmer, M. Apperley, Software for students to sketch interface designs, in: Proceedings of Interact, IOS Press, Amsterdam, 2003, pp. 73–80.

19. B. Plimmer, M. Apperley, Interacting with sketched interface designs: an evaluation study, in: CHI '04: CHI '04 Extended Abstracts on Human factors in Computing Systems, ACM, New York, NY, USA, 2004, pp. 1337–1340, Springer.

20. A. Coyette, J. Vanderdonckt, A sketching tool for designing any user, any platform, anywhere user interfaces, in: INTERACT, 2005, pp. 550–564.

21. M. Fonseca, J. Jorge, Using fuzzy logic to recognize geometric shapes interactively. Fuzzy Systems, 2000. FUZZ IEEE 2000. The Ninth IEEE International Conference, vol. 1, 2000, pp. 291–296.

22. D. Rubine, Specifying gestures by example, SIGGRAPH Comput. Graph. 25 (4) (1991) 329–337.

23. H. Freeman, Computer processing of line-drawing images, ACM Comput. Surv. 6 (1) (1974) 57–97.

24. V.I. Levenshtein, Binary codes capable of correcting deletions, insertions, and reversals, Sov. Phys. Dokl. 10 (1966) (1966) 707–710.

25. H. Oda, A. Kitadai, M. Onuma, M. Nakagawa, A search method for on-line handwritten text employing writing-box-free handwriting recognition, in: IWFHR '04: Proceedings of the Ninth International Workshop on Frontiers in Handwriting Recognition, IEEE Computer Society, Washington, DC, USA, 2004, pp. 545–550.

26. M.P. Perrone, G.F. Russell, A. Ziq, Machine learning in a multimedia document retrieval framework, IBM Syst. J. 41 (3) (2002) 494–503.

27. A.K. Jain, A.M. Namboodiri, Indexing and retrieval of on-line handwritten documents, in: ICDAR '03: Proceedings of the Seventh International Conference on Document Analysis and Recognition, IEEE Computer Society, Washington, DC, USA, 2003, pp. 655–659.

28. D. Lopresti, A. Tomkins, On the searchability of electronic ink, in: Proceedings of the Fourth International Workshop on Frontiers in Handwriting Recognition, 1994, pp. 156–165.

29. J. Makhoul, F. Kubala, R. Schwartz, R. Weischedel, Performance measures for information extraction, in: Proceedings of DARPA Broadcast News Workshop, 1999, pp. 249–252.

Part II

Multimodal Signal Processing and Modelling

Chapter 7

Basic Concepts of Multimodal Analysis

Mihai Gurban and Jean-Philippe Thiran
Ecole Polytechnique Fédérale de Lausanne, Lausanne, Switzerland

7.1 DEFINING MULTIMODALITY

The word *multimodal* is used by researchers in different fields and often with different meanings. One of its most common uses is in the field of human–computer (or man–machine) interaction (HCI). Here, a *modality* is a natural way of interaction: speech, vision, face expressions, handwriting, gestures or even head and body movements. Using several such modalities can lead to *multimodal speaker tracking systems, multimodal person identification systems, multimodal speech recognisers* or, more generally, *multimodal interfaces*. Such interfaces aim to facilitate HCI, augmenting or even replacing the traditional keyboard and mouse. Let us go into more detail about these particular examples.

Multimodal speaker detection, tracking or *localisation* consists of identifying the active speaker in an audio-video sequence which

Multimodal Signal Processing, ISBN: 9780123748256

contains several speakers, based on the correlation between the audio
and the movement in the video [1, 2]. In the case of a system with
several cameras, the speakers might not necessarily be in the same
image, but the underlying idea remains the same. The gain from multimodality over using for example only video is that correlations with
the audio can help discriminate between the person that is actually
speaking and someone who might be only murmuring something
inaudible. It is also possible to find the active speaker using audio
only, with the help of microphone arrays, but this makes the hardware
set-up more complex and expensive. More details on this application
can be found in Chapter 9.

For *multimodal speech recognition*, or *audio-visual speech recognition*, some video information from the speakers' lips is used to
augment the audio stream to improve the speech recognition accuracy [3]. This is motivated by human speech perception, as it has
been proven that humans subconsciously use both audio and visual
information when understanding speech [4]. There are speech sounds
which are very similar in the audio modality but easy to discriminate visually. Using both modalities significantly increases automatic
speech recognition results, especially in the case where the audio is
corrupted by noise. Chapter 9 gives a detailed treatment on how the
multimodal integration is done in this case.

A *multimodal biometrics system* [5] establishes the identity of a
person from a list of candidates previously enrolled in the system,
based on not one but several modalities, which can be taken from a
long list: face images, audio speech, visual speech or lip-reading, fingerprints, iris images, retinal images, handwriting, gait and so on. The
use of more than one modality makes results more reliable, decreasing the number of errors. It should be noted here how heterogenous
these modalities can be.

There are still other examples in the domain of HCI. *Multimodal annotation systems* are annotating systems typically used on
audio-visual sequences, to add metadata describing the speech, gestures, actions and even emotions of the individuals present in these
sequences (see Chapter 12). Indeed, *multimodal emotion recognition* [6] is a very active area of research, using typically the speech,
face expressions, poses and gestures of a person to assess his or her

emotional state. The degree of interest of a person can also be assessed from multimodal data, as shown in Chapter 15.

However, all these applications are taken from just one limited domain, HCI. There are many other fields where different *modalities* are used.

For psychologists, *sensory modalities* represent the human senses (sight, hearing, touch and so on). This is not very different from the previous interpretation. However, other researchers can use the word *modality* in a completely different way. For example, linked to the concept of *medium* as a physical device, *modalities* are the ways to use such *media* [7]. The pen device is a medium, whereas the actions associated to it, such as drawing, pointing, writing or gestures, are all modalities.

However, in signal processing applications, the modalities are signals originating from the same physical source [8] or phenomenon, but captured through different means. In this case, the number of modalities may not be the same as the number of information streams used by the application. For example, the video modality can offer information about the identity of the speaker (through face analysis), his or her emotions (through facial gestures analysis) or even about what he or she is saying (through lip-reading). This separation of information sources in the context of multimodal analysis can be called *information fission*.

Generally, the word *multimodal* is associated to the input of information. However, there are cases where the output of the computer is considered multimodal, as is the case of *multimodal speech synthesis*, the augmentation of synthesised speech with animated talking heads (see Chapter 13 for more details). In this context, distributing the information from the internal states of the system to the different output modalities is done through *multimodal fission*.

In the context of medical image registration, a *modality* can be any of a large array of imaging techniques, ranging from anatomical, such as X-ray, magnetic resonance imaging or ultrasound, to functional, such as functional magnetic resonance imaging or positron emission tomography [9]. *Multimodal registration* is the process of bringing images from several such modalities into spatial alignment. The same term is used in remote sensing, where the modalities are images from different spectral bands [10] (visible, infrared or microwave, for example).

For both medical image analysis and remote sensing, the gain from using images from multiple modalities comes from their complementarity. More information will be present in a combined multimodal image than in each of the monomodal ones, provided they have been properly aligned. This alignment is based on information that is common to all the modalities used, *redundant* information.

7.2 ADVANTAGES OF MULTIMODAL ANALYSIS

Although there is such high variability in multimodal signals, general processing methods exist which can be applied on many problems involving such signals. And the research into such methods is ongoing, as there are important advantages from using multimodal information. The first advantage is the complementarity of the information contained in those signals, that is, each signal brings extra information which can be extracted and used, leading to a better understanding of the phenomenon under study. The second is that multimodal systems are more reliable than monomodal ones because if one modality becomes corrupted by noise, the system can still rely on the other modalities to, at least partially, achieve its purpose. *Multimodal analysis* consists then in exploiting this complementarity and redundancy of multiple modalities to analyse a physical phenomenon.

To offer an example for the complementarity of information, in multimodal medical image analysis, information that is missing in one modality may be clearly visible in another. The same is true for satellite imaging. Here, by missing information, we mean information that exists in the physical reality but was not captured through the particular modality used. By using several different modalities, we can get closer to the underlying phenomenon, which might not be possible to capture with just one type of sensor.

In the domain of human–computer interfaces, combining speech recognition with gesture analysis can lead to new and powerful means of interaction. For example, in a 'put-that-there' interface [11], visual objects can be manipulated on the screen or in a virtual environment with voice commands coupled by pointing. Here, the modalities are naturally complementary because the action is only specified by voice, where the object and place are inferred from the gestures.

Another example could be audio-visual speech recognition. Indeed, some speech sounds are easily confusable in audio but easy to distinguish in video and vice versa. In conditions of noise, humans themselves lip-read subconsciously, and this also shows that there is useful information in the movement of the lips, information that is missing from the corrupted audio.

For the second advantage, the improved reliability of multimodal systems consider the case of a three-modal biometric identification system, based on face, voice and fingerprint identification. When the voice acquisition is not optimal, because of a loud ambient noise or because the subject has a bad cold, the system might still be able to identify the person correctly by relying on the face and fingerprint modalities. Audio-visual speech recognition can also be given as an example because if for some reason the video becomes unavailable, the system should seamlessly revert to audio-only speech recognition.

As can be seen, the extra processing implied by using several different signals is well worth the effort, as both accuracy and reliability can be increased in this way. However, multimodal signal processing also has challenges, of which we will present two of the most important ones. The first is the extraction of relevant features from each modality, features that contain all the information needed at the same time being compact. The second is the fusion of this information, ideally in such a way that variations in the reliability of each modality stream affect the final result as little as possible. We will now analyse in detail why each of these challenges is important.

The first challenge is the extraction of features which are at the same time relevant and compact. By *relevant* features, we mean features that contain the information required to solve the underlying problem, thus the term *relevance* is always tied to the context of the problem. To offer a simple example, for both speech recognition and speaker identification, some features need to be extracted from the audio signal. But their purpose is different. Features extracted for speech recognition should contain as much information as possible about what was being said and as little as possible about who was speaking. Thus, some hypothetic ideal features that would respect this requirement would be very relevant for the speech recognition task but irrelevant for speaker identification. For speaker identification, the situation is reversed, that is, the features should contain as much information as possible about who is speaking and nothing about what was said. Obviously, we would like features to contain all the relevant information in the signal, and at the same time retain as little superfluous information as possible.

The second requirement that we impose on the features is that they should be *compact*, that is, the feature vector should have a low dimensionality. This is needed because of the *curse of dimensionality*, a term defining the fact that the number of equally-distributed samples needed to cover a volume of space grows exponentially with its dimensionality. For classification, accurate models can only be built when an adequate number of samples is available, and that number grows exponentially with the dimensionality. Obviously, for training, only a limited amount of data is available, so having a low input dimensionality is desirable to obtain optimal classification performance.

Dimensionality reduction techniques work by eliminating the *redundancy* in the data, that is, the information that is common to several modalities or, inside a modality, common to several components of the feature vector. The quest for compact features needs, however, to be balanced by the need for *reliability*, that is, the capability of the system to withstand adverse conditions like noise. Indeed, multimodal systems can be tolerant to some degree of noise, in the case when there is redundancy between the modalities. If one of the modalities is corrupted by noise, the same information may be found in another modality. In conclusion, reducing redundancy between the features of a modality is a good idea; however, reducing redundancy between modalities needs to be done without compromising the reliability of the system.

Therefore, for the system to attain optimal performance, the features need to be relevant and compact.

The second challenge that we mentioned is multimodal *integration*, or *fusion*, which represents the method of combining the information coming from the different modalities. As the signals involved may not have the same data rate, range or dimensionality, combining them is not straightforward. The goal here is to gather the useful information from all the modalities, in such a way that the end-result is superior to those from each of the modalities individually. Multimodal fusion can be done at different levels. The simplest kind of fusion operates at the lowest level, the signal level, by concatenating the signals coming from each modality. Obviously, for this, the signals have to be synchronised. High-level decision fusion operates directly on decisions taken on the basis of each modality, and this does not require synchrony. There can also be an intermediate level of fusion, where some initial information is extracted from each modality, synchronously fused into a single stream and then processed further. A more detailed analysis is given in Chapter 8.

The multimodal integration method has an important impact on the reliability of the system. Ideally, the fusion method should allow variations in the importance given to the data streams from each modality because the quality of these streams may vary in time. If one modality becomes corrupted by noise, the system should adapt to rely more on the other modalities, at least until the corrupted modality returns to its previous level of quality. This type of adaptation is, however, not possible with simple low-level fusion but only with decision fusion.

The levels of fusion can be easily illustrated on the example of audio-visual speech recognition. Concatenating the audio and visual features into a single audio-visual data stream would represent the lowest level of fusion. Recognising phonemes and visemes (visual speech units) in each modality and then fusing these speech units into multimodal word models represent intermediate-level fusion. Finally, when recognising the words individually in each modality based on monomodal word models and then fusing only the n-best lists means, fusion was done at the highest level.

Synchrony between the signals is also an important aspect. In some applications, such as audio-visual speech recognition, the signals are naturally synchronous, as phonemes and visemes are emitted simultaneously. In other applications, the signals are asynchronous, such as in the case of simultaneous text and speech recognition. Here, the user is not constrained in pronouncing the words at the same time as they are written, so the alignment needs to be estimated through the analysis. Obviously low-level fusion is not feasible in this case.

7.3 CONCLUSION

We have presented several examples of multimodal applications, showing how, by using the complementarity of different modalities, we can improve classification results, enhance user experience and increase overall system reliability. We have shown the advantages of multimodal analysis but also the main challenges in the field.

The next chapter, Chapter 8, will give a much more detailed overview on the multimodal fusion methods and the levels at which fusion can be done.

REFERENCES

1. H. Nock, G. Iyengar, C. Neti, Speaker localisation using audio-visual synchrony: an empirical study, in: Proceedings of the International Conference on Image and Video Retrieval, pp. 565–570, 2003.
2. M. Gurban, J. Thiran, Multimodal speaker localization in a probabilistic framework, in: Proceedings of the 14th European Signal Processing Conference (EUSIPCO), 2006.
3. G. Potamianos, C. Neti, J. Luettin, I. Matthews, Audio-visual automatic speech recognition: an overview, in: G. Bailly, E. Vatikiotis-Bateson, P. Perrier (Eds.), Issues in Audio-Visual Speech Processing, MIT Press, 2004.
4. H. McGurk, J. MacDonald, Hearing lips and seeing voices, Nature 264 (1976) 746–748.
5. A. Ross, A.K. Jain, Multimodal biometrics: an overview, in: Proceedings of the 12th European Signal Processing Conference (EUSIPCO), pp. 1221–1224, 2004.
6. N. Sebe, I. Cohen, T. Huang, Multimodal emotion recognition, in: C. Chen, P. Wang (Eds.), Handbook of Pattern Recognition and Computer Vision, World Scientific, 2005.
7. C. Benoit, J.C. Martin, C. Pelachaud, L. Schomaker, B. Suhm, Audio visual and multimodal speech systems, in: D. Gibbon (Ed.), Handbook of Standards and Resources for Spoken Language Systems – Supplement Volume, 2000, Mouton de Gruyter, Berlin.
8. T. Butz, J. Thiran, From error probability to information theoretic (multi-modal) signal processing, Signal Process. 85 (2005) 875–902.
9. J.B.A. Maintz, M.A. Viergever, A survey of medical image registration, Med. Image Anal. 2 (1) (1998) 1–36.
10. B. Zitova, J. Flusser, Image registration methods: a survey, Image Vision Comput. 21 (2003) 977–1000.
11. R.A. Bolt, "put-that-there": voice and gesture at the graphics interface, in: SIGGRAPH '80: Proceedings of the 7th Annual Conference on Computer Graphics and Interactive Techniques, vol. 14, ACM Press, 1980, pp. 262–270.

Chapter 8

Multimodal Information Fusion

Norman Poh and Josef Kittler
University of Surrey, UK

8.1 INTRODUCTION

Humans interact with each other using different modalities of communication. These include speech, gestures, documents, etc. It is therefore natural that human–computer interaction (HCI) should facilitate the same multimodal form of communication. To capture this information, one uses different types of sensors, i.e., microphones to capture the audio signal, cameras to capture life video images, 3D sensors to directly capture the surface information, in real time. In each of these cases, commercial off-the-shelf (COTS) devices are already available and can be readily deployed for HCI applications. Examples of HCI applications include audio-visual speech recognition, gesture recognition, emotional recognition and person recognition using biometrics.

Multimodal Signal Processing, ISBN: 9780123748256

The provision of multiple modalities is not only motivated by human factors (i.e., usability, and seamless transition from human–human interaction to HCI) but is also warranted from the engineering point of view. Some of the reasons for not relying on a single mode of communication are given in the following sections:

1. **Noise in sensed data:** One can distinguish three kinds of noise in the sensed data: sensor, channel and modality-specific noise. Sensor noise is the noise that is introduced by the sensor itself. For instance, each pixel in a camera sensor contains one or more light sensitive photo-diodes that convert the incoming light (photons) into an electrical signal. The signal is encoded as the colour value of the pixel of the final image. Even though the same pixel would be exposed several times by the same amount of light, the resulting colour value would not be identical, but have small variation called 'noise'. Channel noise, on the other hand, is the result of degradation introduced by the transmission channel or medium. For example, under slightly changed lighting conditions, the same HCI modality may change. The best known example perhaps is in person recognition, where the same face under changing lighting conditions appear *more differently* than two different faces captured under the same lighting conditions. Finally, modality-specific noise is one that is due to the difference between the captured data and the *canonical* representation of the modality. For instance, it is common to use frontal (mugshot) face to represent the face of a person. As a result, any pose variation in a captured face image can become a potential source of noise.

2. **Non-universality:** A HCI system may not be able to acquire meaningful data from a subset of individuals. For example, a speech recognition system is totally useless to recognise sentences conveyed by a dumb person using a sign language; lip-reading would have some successes. Similarly, in a biometric application, an iris recognition system may be unable to obtain the iris information of a subject with long eyelashes, drooping eyelids or certain pathological conditions of the eye. However, a face recognition system would still be able to provide a useful biometric modality. Clearly, while no single modality is perfect, a combination of them

should ensure wider user coverage, hence improving accessibility, especially to the disabled person.

3. **Upper bound on system performance:** The matching performance of a speech-only-based recognition system cannot be continuously improved by tuning the feature extraction and classifier. There is an implicit upper bound on the number of distinguishable patterns (i.e., the intrinsic upper and lower limits of speech frequency) that can be represented using features derived from spectral envelope. Automatic lip-reading, in this case, may improve the performance further when used in combination with the speech recognition system. In person recognition using fingerprints, for instance, this upper bound is dependent on the biometric feature set, whose capacity is constrained by the variations observable in the feature set of each subject (i.e., *intra*-class variations) and the variations between feature sets of different subjects (i.e., *inter*-class variations). Using multiple biometrics for person recognition has been shown to attain performance beyond that achievable by any single biometric system alone [1] (and references herein).

From the aforementioned arguments, it is obvious that a successful HCI system should rely on multiple modalities. Indeed, apart from audio and visual information as mentioned here, humans also rely on the haptic modality, smell and taste. From these basic sensory information, higher cues such as 3D and temporal information, as well as emotional (e.g., stress, frustration) and psychological state (e.g., interest), can also be derived. Although it may take some time before one can design a general purpose HCI system capable of emulating a human (e.g., a humanoid robot), it is already possible to design a specialised system with some of the functionality needed for effective HCI.

Combining the HCI information that is inherently multimodal introduces new challenges to machine learning. For example, face and speech modalities are often sampled at different rates (30 frames per second vs. 44K samples per second), each often have different representations (2D vs. 1D in this case), and as a result, often are processed with different machine-learning algorithms (see Chapter 7). This chapter focuses on issues raised by combining multiple modalities in HCI systems. Combining several systems has been investigated in pattern recognition [2] in general; in applications related to audio-visual speech processing [3–5]; in speech recognition – examples of methods are multiband [6], multistream [7, 8], front-end multi-feature [9] approaches and the union model [10]; in the form of ensemble [11]; in audio-visual person authentication [12]; and in multibiometrics [1, 13–16] (and references herein), among others. In fact, for audio-visual person authentication, one of the earliest works addressing multimodal biometric fusion was reported in 1978 [17]. Therefore, multimodal fusion has a history of 30 years.

This chapter is organised as follows: Section 8.2 presents different levels of information fusion. Section 8.3 presents a relatively recent type of fusion utilising modality-dependent signal quality. Section 8.4 discusses issues related to designing a fusion classifier. Finally, Section 8.5 concludes the chapter.

8.2 LEVELS OF FUSION

On the basis of the type of information available in a certain module, different levels of fusion may be defined. Sanderson and Paliwal [18] categorise the various levels of fusion into two broad categories: pre-classification or fusion *before* matching, and post-classification or fusion *after* matching (Figure 8.1). The latter has been attracting a lot of attention because the amount of information available for fusion reduces drastically once the matcher has been invoked. Pre-classification fusion schemes typically require the development of new matching techniques (as the matchers/classifiers used by the individual sources may no longer be relevant) thereby introducing additional challenges. Pre-classification schemes include fusion at the sensor (or raw data) and the feature levels while post-classification schemes include fusion at the match score, rank and decision levels (see also [1]).

1. **Sensor-level fusion:** The raw data (e.g., a face image) acquired from an individual represents the richest source of information although it is expected to be contaminated by noise (e.g., non-uniform illumination, background clutter, etc.). Sensor-level

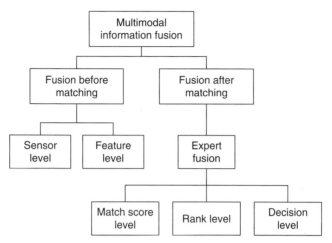

FIGURE 8.1   Various levels of multimodal information fusion can be accomplished at various levels in a biometric system.

fusion refers to the consolidation of (a) raw data obtained using multiple sensors, or (b) multiple snapshots of a biometric using a single sensor [19, 20].

2. **Feature-level fusion:** In feature-level fusion, the feature sets originating from multiple feature extraction algorithms are consolidated into a single feature set by the application of appropriate feature normalisation, transformation and reduction schemes. The primary benefit of feature-level fusion is the detection of correlated feature values generated by different feature extraction algorithms and, in the process, identifying a salient set of features that can improve recognition accuracy. Eliciting this feature set typically requires the use of dimensionality reduction/selection methods and, therefore, feature-level fusion assumes the availability of a large number of training data. Also, the feature sets being fused are typically expected to reside in commensurate vector spaces such that a matching technique can be applied easily in order to match two patterns [21, 22].

3. **Score-level fusion:** At this level, the match scores output by multiple experts are combined to generate a new output (a scalar or vector) that can be subsequently used for decision-making. Fusion at this level is the most commonly discussed approach primarily due

to the ease of accessing and processing match scores (compared to the raw data or the feature set extracted from the data). Fusion methods at this level can be broadly classified into three categories [1]: density-based schemes [23,24] (generative approach), classifier-based schemes (discriminative approach) [25] and transformation-based schemes [26] (see also Section 8.4).

4. **Rank-level fusion:** In HCI applications, the output of the system can be viewed as a ranking of plausible hypotheses. In other words, the output indicates the set of possible hypotheses sorted in decreasing order of confidence. The goal of rank level fusion schemes is to consolidate the ranks output by the individual expert systems to derive a consensus rank for each hypothesis. Ranks provide more insight into the decision-making process of the matching compared to just the best hypothesis, but they reveal less information than match scores. However, unlike match scores, the rankings output by multimodal experts are comparable. As a result, no normalisation is needed and this makes rank level fusion schemes simpler to implement compared to the score level fusion techniques [2].

5. **Decision-level fusion:** Many commercial off-the shelf (COTS) HCI systems provide access only to the final recognition decision. When such COTS devices are used to build a multimodal system, only decision level fusion is feasible. Methods proposed in the literature for decision level fusion include 'AND' and 'OR' rules [27], majority voting [28], weighted majority voting [29], Bayesian decision fusion [30], the Dempster-Shafer theory of evidence [30] and behaviour knowledge space [31].

8.3 ADAPTIVE VERSUS NON-ADAPTIVE FUSION

One can also distinguish fusion classifiers by whether they are adaptive or non-adaptive. Adaptive, or quality-based fusion attempts to change the weight associated with a modality as a function of the signal quality measured on the modality. The idea is to give higher weights to the modality with higher quality. For instance, in biometric person recognition, if the facial image is corrupted by bad illumination, the output of the speech system may be weighed more, and vice-versa when the speech system is corrupted by noise.

The quality of an incoming modality is measured by quality measures. *Quality measures* are a set of criteria designed to assess the quality of the incoming signal of a modality. Examples of quality measures for face images are face detection reliability, presence of glasses, brightness, contrast, etc. [32]. For speech, this would be signal-to-noise ratio and speech-likeness (vs. noise). An ideal quality measure should correlate, to some extent, with the performance of the classifier processing the modality [33]. For instance, if a face recognition system can degrade in performance due to change of head pose, then head pose is an ideal quality measure candidate. In practice, more than one quality measure is often needed as each measure can only quantify one particular aspect of the signal quality; and a pool of measurements will, in principle, gauge as many degrading factors as possible. Figure 8.2 shows the difference between an adaptive and a non-adaptive fusion scheme.

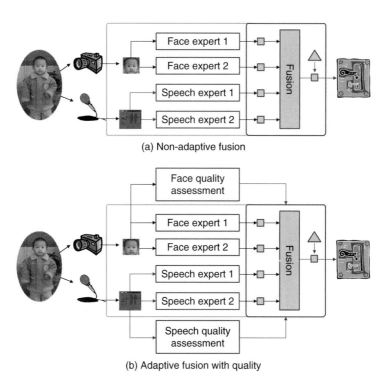

(a) Non-adaptive fusion

(b) Adaptive fusion with quality

FIGURE 8.2   Adaptive versus non-adaptive fusion.

One can show that such a fusion strategy requires a non-linear solution in the underlying system output space. Let $y_{\text{com}} \in \mathbb{R}$ be the combined score and $y_i \in \mathbb{R}$ be the output of each modality (after matching; hence producing, for instance, posterior probability of a class label). A non-adaptive linear fusion classifier will take the form of:

$$y_{\text{com}} = \sum_{i=1}^{N} w_i y_i + w_0$$

where $w_i \in \mathbb{R}$ is the weight associated to the output y_i and w_0 is a bias term[1]. In contrast, the adaptive fusion classifier would compute

$$y_{\text{com}} = \sum_{i=1}^{N} w_i(q) y_i + w_0(q) \tag{8.1}$$

where $w_i(q)$ changes with the quality signal q. For the sake of simplicity, we shall assume for now that $q = \{q_1, \ldots, q_N\}$, where q_i is the quality measure of the i-th modality. In general $w_i(q)$ could be of any functional form. However, we shall assume that weights vary linearly as a function of quality, i.e.,

$$w_i(q) = \sum_i w_i^{(2)} q_i + w_i^{(1)} \tag{8.2}$$

and

$$w_0(q) = \sum_i w_i^{(0)} q_i + w_0^{(0)} \tag{8.3}$$

---

1. Essentially, y_{com} is the output of a *discriminant function*. To consider y_{com} as probability, one can use a logistic or sigmoid output instead, i.e., $y_{\text{com}} = \frac{1}{1+\exp(-a)}$, where $a = \sum_{i=1}^{N} w_i y_i + w_0$. Generalisation to multiple-class hypotheses is straightforward, as discussed in [34, 35].

Substituting (8.2) and (8.3) into (8.1), and rearranging, we find:

$$y_{com} = \sum_{i=1}^{N} y_i \left(w_i^{(2)} q_i + w_i^{(1)} \right) + \left(\sum_{i=1}^{N} w_i^{(0)} q_i + w_0^{(0)} \right)$$

$$= \sum_{i=1}^{N} w_i^{(2)} y_i q_i + \sum_{i=1}^{N} w_i^{(1)} y_i + \sum_{i=1}^{N} w_i^{(0)} q_i + w_0^{(0)} \qquad (8.4)$$

where we note that $w_i^{(2)}$ is the weight associated to the pairwise element $y_i \cdot q_i$, $w_i^{(1)}$ is the weight associated to y_i and $w_i^{(0)}$ is the weight associated to q_i.

From (8.4), it is obvious that one way to realise (8.1) is to consider a classifier taking $\{y_i q_i, y_i, q_i\}$ as features.

The adaptive fusion can be seen as a function $\{y_i, q_i | i = 1, \ldots, N\} \rightarrow y_{com}$ taking the product of y_i and q_i into consideration. This shows that adaptive fusion, in the simplest case, is a linear function of quality, and such a function is non-linear in the space of $\{y_i | i = 1, \ldots, N\}$.

There are two ways quality measures can be incorporated into a fusion classifier, depending on their role, i.e., either as a control parameter or as evidence. In their primary role, quality measures are used to modify the way a fusion classifier is trained or tested, as suggested in the Bayesian-based classifier called 'expert conciliation' [36], reduced polynomial classifier [37], quality-controlled support vector machines [38] and quality-based fixed rule fusion [39]. In their secondary role, quality measures are often concatenated with the expert outputs to be fed to a fusion classifier, as found in logistic regression [40] and the mixture of Gaussian Bayesian classifiers [41].

Other notable work includes the use of Bayesian networks to gauge the complex relationship between expert outputs and quality measures, e.g., Maurer and Baker's Bayesian network [42] and Poh et al.'s quality state-dependent fusion [32]. The work in [32] takes into account an array of quality measures rather than representing quality as a scalar. By means of grouping the multifaceted quality measures, a fusion strategy can then be devised for each cluster of quality values.

Other suggestions include the use of quality measures to improve HCI device interoperability [43,44]. Such an approach is commonly used in speaker verification [45] where different strategies are used for different microphone types.

Last but not least, another promising direction in fusion is to consider the reliability estimate of each biometric modality. In [46], the estimated reliability for each biometric modality was used for combining decision-level decisions, whereas in [47–50], score-level fusion was considered. However, in [47–49], the term 'failure prediction' was used instead. Such information, derived solely from the expert outputs (instead of quality measures), has been demonstrated to be effective for single modalities [47], fusion across sensors for a single modality [48] and across different machine-learning techniques [49]. In [50], the notion of reliability was captured by the concept of margin, a concept used in large-margin classifiers [51]. Exactly how the reliability is defined and estimated for each modality, and how it can be effectively used in fusion, are still open research issues.

8.4   OTHER DESIGN ISSUES

In the previous section, we highlighted the need of designing an effective fusion mechanism and a set of quality measures *dependent* on the modality and its underlying classifier. Apart from these two, there are also the following issues:

- **Fusion strategies:** An important consideration when adopting a fusion strategy is to consider the statistical dependency among the expert outputs. For instance, in intramodal fusion, several experts may rely on the same biometric sample and so higher dependency is expected among the expert outputs. On the other hand, in a multimodal setting, the pool of experts is likely to be statistically independent (Figure 8.4). In [40], three types of frameworks are proposed to solve a multimodal fusion problem involving intramodal experts. The first framework simply assumes independence, in which case the fusion classifier reduces to a Naive Bayes one. In this case, the output of each expert can be transformed to the same domain (e.g., probability) and the transformed outputs can be combined using simple rules such as sum or product

(Figure 8.5a). The second framework considers dependency of experts in an intramodal setting (all observing the same information) whereas ignores the dependency at the multimodal setting, hence realising a two-stage fusion process (Figure 8.5c). Note that the second stage of fusion in this case can be based on the Naive Bayes principle (using simple product rule, but can also be approximated using the sum rule). Finally, the third framework makes no assumption about the expert outputs (Figure 8.5b). A comparison of these fusion strategies for biometric person authentication can be found in [40].

- **Expert output transformation:** Very often, the outputs of different experts are not comparable. Transforming the outputs into a common domain may help improve the ease of design of a fusion classifier. One has the choice of transforming into probability or log-likelihood ratio domain [14]. In the latter case, the output distribution tends to be normal, hence making multivariate analysis such as correlation more reliable. Figure 8.3 illustrates the transformation of the output of two multimodal experts into log-likelihood ratio space.

- **Choice of architecture:** There is a huge space of different fusion architectures that has not been explored. The range of possible configurations encompassing serial, parallel and hybrid structures is immense. Although the parallel fusion strategy is most commonly

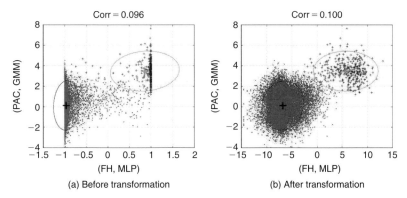

(a) Before transformation   (b) After transformation

FIGURE 8.3   Transformation of heterogeneous expert output to log-likelihood ratio.

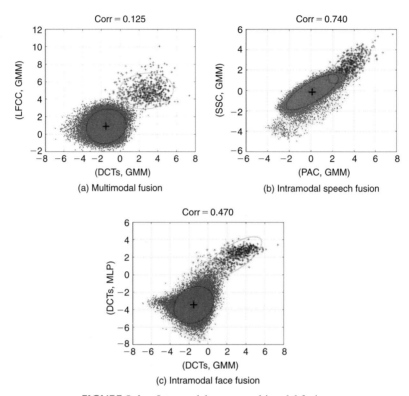

FIGURE 8.4   Intramodal versus multimodal fusion.

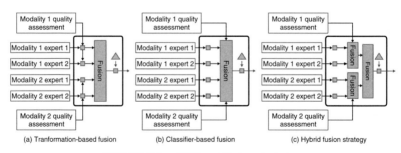

FIGURE 8.5   Different fusion strategies.

used in multimodal HCI information fusion, there are additional advantages in exploring serial fusion, where the experts are considered one at a time. It offers the possibility of making reliable decisions with only a few experts, leaving only difficult samples to be handled by the remaining experts.

8.5   CONCLUSIONS

This chapter gives an overview of multimodal information fusion from the machine-learning perspective. Although information can be combined at different levels (data, feature, score, rank and decision), score-level fusion offers the best trade-off between information complexity and the flexibility in modelling the dependency among different information sources. In particular, we presented a relatively recent fusion approach to multimodal fusion, namely, adaptive fusion, where the idea is to measure the signal quality of each modality and then use this information at the fusion level. In this way, the modality expert with better signal quality impacts on the final decision more heavily. Research in adaptive multimodal fusion is gaining momentum. There are still open problems related to its applications. Among others are

- methodology for designing an effective set of modality-specific and/or classifier-specific quality measures;
- methodology for designing an optimal adaptive (quality-based) fusion mechanism;
- mechanisms for combining expert opinions regarding the meaning of the communicated information.

REFERENCES

1. A. Ross, K. Nandakumar, A.K. Jain, Handbook of Multibiometrics, 1st ed., Springer, New York, USA, 2006.
2. T.K. Ho, J.J. Hull, S.N. Srihari, Decision combination in multiple classifier systems, IEEE Trans. Pattern Anal. Mach. Intell. 16 (1) (1994) 66–75.
3. S. Lucey, Audio Visual Speech Processing, PhD thesis, Queensland University of Technology, 2002.

4. J. Luettin, Visual Speech and Speaker Recognition, PhD thesis, Department of Computer Science, University of Sheffield, 1997.

5. T. Chen, R. Rao, Audio-visual integration in multimodal communications, Proc. IEEE, 86 (5) (1998) 837–852.

6. C. Cerisara, Contribution de l'Approache Multi-Bande à la Reconnaissance Automatic de la Parole, PhD thesis, Institute Nationale Polytéchnique de Lorraine, Nancy, France, 1999.

7. S. Dupont, Étude et Dévelopment de Nouveaux Paradigmes pour la Reconnaissance Robuste de la Parole, PhD thesis, Laboratoire TCTS, Universite de Mons, Belgium, 2000.

8. A. Hagen, Robust Speech Recognition Based on Multi-Stream Processing, PhD thesis, Ecole Polytechnique Federale de Lausanne, Switzerland, 2001.

9. M.L. Shire, Discriminant Training of Front-End and Acoustic Modelling Stages to Heterogeneous Acoustic Environments for Multi-Stream Automatic Speech Recognition, PhD thesis, University of California, Berkeley, USA, 2001.

10. J. Ming, F.J. Smith, Speech recognition with unknown partial feature corruption – a review of the union model. Comput. Speech Lang. 17 (2003) 287–305.

11. G. Brown, Diversity in Neural Network Ensembles, PhD thesis, School of Computer Science, University of Birmingham, 2003.

12. C. Sanderson, Automatic Person Verification Using Speech and Face Information, PhD thesis, Griffith University, Queensland, Australia, 2002.

13. K. Nandakumar, Integration of Multiple Cues in Biometric Systems, Master's thesis, Michigan State University, 2005.

14. N. Poh, Multi-system Biometric Authentication: Optimal Fusion and User-Specific Information, PhD thesis, Swiss Federal Institute of Technology in Lausanne (Ecole Polytechnique Fédérale de Lausanne), 2006.

15. K. Kryszczuk, Classification with Class-independent Quality Information for Biometric Verification, PhD thesis, Swiss Federal Institute of Technology in Lausanne (Ecole Polytechnique Fédérale de Lausanne), 2007.

16. J. Richiardi, Probabilistic Models for Multi-Classifier Biometric Authentication Using Quality Measures, PhD thesis, Swiss Federal Institute of Technology in Lausanne (Ecole Polytechnique Fédérale de Lausanne), 2007.

17. A. Fejfar, Combining techniques to improve security in automated entry control, in: Carnahan Conf. On Crime Countermeasures, Mitre Corp. MTP-191, 1978.

18. C. Sanderson, K.K. Paliwal, Information Fusion and Person Verification Using Speech and Face Information. Research Paper IDIAP-RR02-33, IDIAP, September 2002.

19. R. Singh, M. Vatsa, A. Ross, A. Noore, Performance enhancement of 2D face recognition via mosaicing, in: Proceedings of the 4th IEEE Workshop on

Automatic Identification Advanced Technologies (AuotID), Buffalo, USA, 2005, pp. 63–68.

20. A. Ross, S. Shah, J. Shah, Image versus feature mosaicing: a case study in fingerprints, in: Proceedings of SPIE Conference on Biometric Technology for Human Identification III, Orlando, USA, 2006, pp. 620208-1–620208-12.

21. A. Ross, R. Govindarajan, Feature level fusion using hand and face biometrics, in: Proceedings of SPIE Conference on Biometric Technology for Human Identification II, vol. 5779, Orlando, USA, 2005, pp. 196–204.

22. B. Son, Y. Lee, Biometric authentication system using reduced joint feature vector of iris and face, in: Proceedings of Fifth International Conference on Audio- and Video-Based Biometric Person Authentication (AVBPA), Rye Brook, USA, 2005, pp. 513–522.

23. S.C. Dass, K. Nandakumar, A.K. Jain, A principled approach to score level fusion in multimodal biometric systems, in: Proceedings of Fifth International Conference on Audio- and Video-based Biometric Person Authentication (AVBPA), Rye Brook, USA, 2005, pp. 1049–1058.

24. B. Ulery, A. Hicklin, C. Watson, W. Fellner, P. Hallinan, Studies of Biometric Fusion. Technical Report NISTIR 7346, NIST, September 2006.

25. P. Verlinde, G. Cholet, Comparing decision fusion paradigms using k-NN based classifiers, decision trees and logistic regression in a multi-modal identity verification application, in: Proceedings of Second International Conference on Audio- and Video-Based Biometric Person Authentication (AVBPA), Washington D.C., USA, 1999, pp. 188–193.

26. A.K. Jain, K. Nandakumar, A. Ross, Score normalization in multimodal biometric systems, Pattern Recognit. 38 (12) (2005) 2270–2285.

27. J. Daugman, Combining multiple biometrics, <http://www.cl.cam.ac.uk/users/jgd1000>, 2000.

28. L. Lam, C.Y. Suen, Application of majority voting to pattern recognition: an analysis of its behavior and performance, IEEE Trans. Syst. Man. Cybern. A Syst. Hum. 27 (5) (1997) 553–568.

29. L.I. Kuncheva, Combining Pattern Classifiers – Methods and Algorithms. Wiley, 2004.

30. L. Xu, A. Krzyzak, C.Y. Suen, Methods for combining multiple classifiers and their applications to handwriting recognition, IEEE Trans. Syst. Man. Cybern. 22 (3) 1992, 418–435.

31. Y.S. Huang, C.Y. Suen, Method of combining multiple experts for the recognition of unconstrained handwritten numerals, IEEE Trans. Pattern Anal. Mach. Intell. 17 (1) (1995) 90–94.

32. N. Poh, G. Heusch, J. Kittler, On combination of face authentication experts by a mixture of quality dependent fusion classifiers, in: LNCS 4472, Multiple Classifiers System (MCS), Prague, 2007, pp. 344–356.

33. K. Kryszczuk, A. Drygajlo, Improving classification with class-independent quality measures: q-stack in face verification, in: LNCS 4472, Multiple Classifiers System (MCS), 2007, pp. 1124–1133.

34. R.O. Duda, P.E. Hart, D.G. Stork, Pattern Classification and Scene Analysis, John Wiley and Sons, New York, 2001.

35. C. Bishop, Neural Networks for Pattern Recognition, Oxford University Press, 1999.

36. J. Bigun, J. Fierrez-Aguilar, J. Ortega-Garcia, J. Gonzalez-Rodriguez, Multi-modal biometric authentication using quality signals in mobile communications, in: 12th Int'l Conf. on Image Analysis and Processing, Mantova, 2003, pp. 2–13.

37. K.-A. Toh, W.-Y. Yau, E. Lim, L. Chen, C.-H. Ng, Fusion of auxiliary information for multimodal biometric authentication, in: LNCS 3072, Int'l Conf. on Biometric Authentication (ICBA), Hong Kong, 2004, pp. 678–685.

38. J. Fierrez-Aguilar, J. Ortega-Garcia, J. Gonzalez-Rodriguez, J. Bigun, Kernel-based multimodal biometric verification using quality signals, in: Defense and Security Symposium, Workshop on Biometric Technology for Human Identification, Proceedings of SPIE, vol. 5404, 2004, pp. 544–554.

39. O. Fatukasi, J. Kittler, N. Poh, Quality controlled multimodal fusion of biometric experts, in: 12th Iberoamerican Congress on Pattern Recognition CIARP, Viña del Mar-Valparaiso, Chile, 2007, pp. 881–890.

40. J. Kittler, N. Poh, O. Fatukasi, K. Messer, K. Kryszczuk, J. Richiardi, et al., Quality dependent fusion of intramodal and multimodal biometric experts, in: Proceedings of SPIE Defense and Security Symposium, Workshop on Biometric Technology for Human Identification, vol. 6539, 2007.

41. K. Nandakumar, Y. Chen, S.C. Dass, A.K. Jain, Likelihood ratio based biometric score fusion, IEEE Trans. Pattern Anal. Mach. Intell. 30 (2008) 342–347.

42. D.E. Maurer, J.P. Baker, Fusing multimodal biometrics with quality estimates via a Bayesian belief network, Pattern Recognit. 41 (3) (2007) 821–832.

43. F. Alonso-Fernandez, J. Fierrez, D. Ramos, J. Ortega-Garcia, Dealing with sensor interoperability in multi-biometrics: The upm experience at the biosecure multimodal evaluation 2007, in: Proceedings of SPIE Defense and Security Symposium, Workshop on Biometric Technology for Human Identification, 2008.

44. N. Poh, T. Bourlai, J. Kittler, Improving biometric device interoperability by likelihood ratio-based quality dependent score normalization, in: IEEE Conference on Biometrics: Theory, Applications and Systems, Washington, D.C., 2007, pp. 1–5, in press.

45. R. Auckenthaler, M. Carey, H. Lloyd-Thomas, Score normalization for text-independant speaker verification systems, Digit. Signal Process. (DSP) J. 10 (2000) 42–54.

46. K. Kryszczuk, J. Richiardi, P. Prodanov, A. Drygajlo, Reliability-based decision fusion in multimodal biometric verification systems, EURASIP J. Adv. Signal Process. 2007.

47. W. Li, X. Gao, T. Boult, Predicting biometric system failure, in: Computational Intelligence for Homeland Security and Personal Safety, 2005. CIHSPS 2005. Proceedings of the 2005 IEEE International Conference, 2005, pp. 57–64.

48. B. Xie, T. Boult, V. Ramesh, Y. Zhu, Multi-camera face recognition by reliability-based selection, in: Computational Intelligence for Homeland Security and Personal Safety, Proceedings of the 2006 IEEE International Conference, 2006, pp. 18–23.

49. T.P. Riopka, T.E. Boult, Classification enhancement via biometric pattern perturbation, in: AVBPA, 2005, pp. 850–859.

50. N. Poh, S. Bengio, Improving fusion with margin-derived confidence in biometric authentication tasks, in: LNCS 3546, 5th Int'l. Conf. Audio- and Video-Based Biometric Person Authentication (AVBPA 2005), New York, 2005, pp. 474–483.

51. A.J. Smola, P.J. Bartlett (Eds.), Advances in Large Margin Classifiers, MIT Press, Cambridge, MA, 2000.

Chapter 9

Modality Integration Methods

Mihai Gurban and Jean-Philippe Thiran

Ecole Polytechnique Fédérale de Lausanne, Lausanne, Switzerland

9.1 INTRODUCTION

In Chapter 7, the basic concepts of multimodal analysis have been presented, together with a few general examples. Chapter 8 showed how information from different modalities can be fused in a human computer interaction (HCI) system, presenting the different levels and types of fusion. In this chapter, we will present how multimodal fusion can be implemented in practice, on two concrete examples, and how multimodal analysis can lead to better results than just using monomodal information.

The two example applications that we chose are audio-visual speech recognition (AVSR) and multimodal speaker localisation.

Multimodal Signal Processing, ISBN: 9780123748256

The first application has been chosen, as it is a very good example especially for high-level decision fusion methods, whereas the second is better suited for lower-level ones.

Although we present here fusion methods in the context of two specific applications, as we want to emphasise the practical not the theoretical aspects of the methods, the applicability of these algorithms is not limited to the particular applications presented. Indeed, most of the methods shown are general and do not depend on any specific attribute of the modalities involved (audio and video) or the information sought (in our case, speech).

9.2 MULTIMODAL FUSION FOR AVSR

9.2.1 Types of Fusion

AVSR [1] is a multimodal classification problem involving time-varying signals from two modalities, audio and video, which have different rates and properties. For example, they have different temporal resolutions, audio having a sampling rate of tens of kilohertz, whereas video has thousands of times less, that is, only tens of temporal samples per second. They also have a different dimensionality, as video has two spatial dimensions and a temporal one, whereas audio only has a temporal dimension.

AVSR uses visual information derived from the video of the speaker, in particular from the mouth region, to improve the audio speech recognition results, especially when the audio is corrupted by noise. This can be done because the audio and the video are complementary in this case, that is, the phonemes that are easily confused in the audio modality are more distinguishable in the video one and vice versa. Obviously the audio remains the dominant modality in this case, and only slight improvements are obtained from multimodal analysis when the audio is clean; however, the improvements become larger and larger as the audio becomes more corrupted.

Just like in audio-only speech recognition, hidden Markov models (HMMs) [2, 3], which have been presented in Chapter 2, are the dominant classifiers used in virtually all AVSR systems. In small vocabulary systems, each word can be modelled by an HMM, whereas in large vocabulary tasks, each speech sound has a corresponding HMM. Most recognition systems use left–right models, or Bakis models [4], in which states that have been left are never visited again and the system always progresses from an initial state towards a final one.

As mentioned in the introduction, the best-suited methods of multimodal fusion for AVSR are the decision-level methods, and the reason for this is that they allow the modelling of the reliability of each modality, depending on the changing conditions in the audio-visual environment. However, it is still possible to use low-level feature fusion methods such as *feature concatenation* [5], where the audio and video feature vectors are simply concatenated before being presented to the classifier. Here, a single classifier is trained with combined data from the two modalities, and it would be very difficult to weigh the modalities according to their reliability.

As opposed to the feature fusion methods, *decision fusion* or *fusion after matching* methods, as referred to in the previous chapter, allow the assignment of weights to each modality. In this method, separate audio and video classifiers are trained, and their output scores are combined with appropriate weights. There are three possible levels for combining individual modality scores [1]:

- Early integration: HMM state likelihoods are combined, forcing stream synchrony.
- Late integration: N-best hypothesis are combined, from separate audio and visual HMMs.
- Intermediate integration: Models force synchrony at the phone or word boundaries.

To draw a parallel with the previous chapter, early integration is fusion done at the score level, where individual scores are given for each state of the HMM. Late integration is equivalent to decision-level fusion, because the only information available from each modality is in the N-best lists of recognised words. Finally, intermediate fusion is still score-level fusion but done at a larger scale, on groups of HMM states, groups which can be either phonemes or words.

In the following, we will present the early integration strategy which is maybe the most common in AVSR and also the simplest.

9.2.2 Multistream HMMs

Multistream HMMs (MSHMMs) are models which belong to the early integration category, forcing synchrony at the frame level. This type of integration allows very rapid changes in the importance given to each modality, allowing the implementation of systems which can very quickly adapt to changing conditions.

The MSHMM is a statistical model derived from the HMM and adapted for multimodal processing. Unlike typical HMMs which have one Gaussian mixture model (GMM) per state, the MSHMM has several GMMs per state, one for each input modality.

The emission likelihood b_j for state j and observation o_t at time t is the product of likelihoods from each modality s weighted by stream exponents λ_s [6]:

$$b_j(o_t) = \prod_{s=1}^{S} \left[\sum_{m=1}^{M_s} c_{jsm} N(o_{st}; \mu_{jsm}, \Sigma_{jsm}) \right]^{\lambda_s}, \qquad (9.1)$$

where $N(o; \mu, \Sigma)$ is the value in o of a multivariate Gaussian with mean μ and covariance matrix Σ. M_s Gaussians are used in a mixture, each weighed by c_{jsm}. The product in equation (9.1) is in fact equivalent to a weighted sum in logarithmic domain. In practice, the weights λ_s should be tied to stream reliability such that when environment conditions [e.g., signal-to-noise ratio (SNR)] change, they can be adjusted to emphasise the most reliable modality. In the following, we present some common stream reliability estimation methods.

9.2.3 Stream Reliability Estimates

Going back to the concepts of *adaptive* and *nonadaptive fusion* presented in the previous chapter, the weights for AVSR need to be adaptive, that is, they need to reflect the relative quality of the two modalities. However, this adaptation can either be done once, fixing the weights to constant values, or can be dynamic, being redone continuously as the conditions in the environment change.

The choice between constant and dynamic weights is taken in the beginning based on the assumptions made on the context of the problem. If it is assumed that the acoustic and visual environment remains more or less the same and that the training conditions reasonably match those in testing, fixed weights can be used. However, it is more realistic to assume that conditions will not be matched and that they will not be constant in time. Indeed, sudden bursts of noise can happen anytime in the audio, such as cough, a pop from a microphone, the door of the conference room that is swung against the wall or any other similar situation can lead to a sudden degradation in the quality of the audio. Similarly, for the video, a lighting change, a speaker who turns his head or gestures with his hands making his mouth invisible to the camera, all such situations can lead to sudden degradation in the quality of the video stream. These degradations can be temporary as in the case of the cough, or permanent, as the lighting changes. In all these conditions, having the stream weights adapt automatically to the conditions which change in time should be beneficial for the performance of the system.

Fixed Stream Weights

Fixed stream weights can be derived with discriminative training techniques, applied on training or held-out data. They will only be relevant for the particular environment conditions in which that data were acquired. From the methods that are applied directly on the training data, some minimise a smooth function of the word classification error [7, 8]. Another approach is to minimise the frame classification error, as in [9] where the maximum entropy criterion is used. From the methods that use a held-out data set, the simplest is the grid search, when the weights are constant and constrained to sum to 1, as is the case in [9] or [10]. More complex methods need to be used in other cases, for example, when weights are class-dependent; however, this dependency was not proved to have a beneficial effect on recognition results, as shown in [9] or [11].

Yet another approach is to use a small amount of unlabelled data, as in [12], to estimate the stream weights in an unsupervised way. Class specific models and anti-models are first built and then used to initialise a k-means algorithm on the unlabelled data. The stream weights ratio is then expressed as a nonlinear function of intra- and inter-class distances.

In [13], the testing set itself is used in an iterative likelihood-ratio maximisation algorithm to determine the stream weights. The algorithm finds the weights that maximise the dispersion of stream emission likelihoods $P(o|c)$, which should lead to better classification results. The measure to be maximised is

$$L(\lambda_c^A) = \sum_{t=1}^{T} \sum_{c \in C} \left\{ P(o_t^A | c_t) - P(o_t^A | c) \right\}, \qquad (9.2)$$

where c is the class or HMM state out of the set C of classes, and o^A is the audio observation vector. The measure is computed over a time interval of T frames.

An extension of this algorithm is based on output likelihood normalisation [14]. Here, the weights are class-dependent, and the weight for one class is the ratio between the average class-likelihood for a time period T over the N classes and the average likelihood for that particular class over the same time period, that is,

$$l_{vt}^A = \frac{\frac{1}{NT} \sum_{t=1}^{T} \sum_{c \in C} \log P(o_t^A | c)}{\frac{1}{T} \sum_{t=1}^{T} \log P(o_t^A | v)}. \qquad (9.3)$$

Both these methods optimise the audio weights first and then set the video weights relative to the audio ones.

Stream reliability estimates are however not limited to the AVSR field. For example, in [15], reliability measures are used for the fusion of three streams for multimodal biometric identification. The three streams are audio speech, visual speech and the speaker's face, while the reliability measure used is the difference between the two highest ranked scores, normalised by the mean score.

Dynamical Stream Weights

Dynamical stream weights are however better suited for practical systems, where the noise can vary unexpectedly. Such examples of sudden degradation of one modality can be loud noises in the audio or loss of face/mouth tracking in the video. Events like these can happen in a practical set-up, and they prove the need for temporally-varying stream weights. The weights can be adjusted for example based on the estimated SNR, as in [5, 16, 17, 18, 19, 20, 21], or based on the voicing index [22] used in [23]. However, these methods are based on reliability measures on the audio only, and the video degradation is not taken into account. Other weighting methods are modality-independent, based only on indicators of classifier confidence, as presented in the following.

In [24] and [25], several classifier confidence measures are used. The first one is the N-best log-likelihood difference, based on the stream emission likelihoods $P(o|c)$. The measure is defined as follows. If o_{st} is the observation for stream s at time t and c_{stn} are the N-best most likely generative classes (HMM states), $n = 1...N$, then the log-likelihood difference at time t for stream s is

$$l_{st} = \frac{1}{N-1} \sum_{n=2}^{N} \log \frac{P(o_{st}|c_{st1})}{P(o_{st}|c_{stn})}. \tag{9.4}$$

The justification is that the likelihood ratios give a measure of the reliability of the classification. The same justification can be given for the second measure used, the log-likelihood dispersion, defined as

$$d_{st} = \frac{2}{N(N-1)} \sum_{m=1}^{N} \sum_{n=m+1}^{N} \log \frac{P(o_{st}|c_{stm})}{P(o_{st}|c_{stn})}. \tag{9.5}$$

Other methods for stream confidence estimation are based on posteriors not on likelihoods. In [26], the class-posterior probability of the combined stream $P(c|o^{AV})$ is computed as the maximum between three posteriors, derived from three observation vectors, the audio-only one o_t^A, the video-only o_t^V or the concatenated observation o_t^{AV}, that is,

$$P(c_{tn}|o_t) = \max \left(P\left(c_{tn}|o_t^A\right), P\left(c_{tn}|o_t^V\right), P\left(c_{tn}|o_t^{AV}\right) \right). \tag{9.6}$$

The stream reliability estimation framework is not only applicable on AVSR but also in audio-only speech recognition, in the case when multiple feature streams are used to exploit their complementarity. For example, in [27], the entropy of the class-posterior distribution is used as a reliability estimator:

$$h_{st} = -\sum_{i=1}^{C} P(c_i|o_{st}) \log P(c_i|o_{st}), \tag{9.7}$$

where C is the number of classes. This criterion is similar to the dispersion of the N-best hypotheses, presented above; however, it is applied on the distribution of all the classes not just the best N ones.

This last method has also been used in AVSR [28], to compute the stream weights λ_s based on the estimated stream reliability, which is derived from the entropy of the class-posterior distributions for each stream.

The reasoning is as follows. When the class-posterior distribution has a clear peak, there is a very good match between the test sample and the class model for the recognised class and a very bad match with all the other classes. The confidence that we have in assigning the sample to the class is high, meaning that the confidence in the corresponding stream should also be high. On the other hand, when the posterior distribution is flat or nearly flat, there is a high possibility that the sample was assigned to the wrong class, so the confidence is low, both in the classification result and the stream. Entropy is a good measure of the form of the probability distribution because it is high when the distribution is flat and low when it has a clear peak.

The big advantage of this algorithm is its flexibility in different environments. If one of the modalities becomes corrupted by noise, the posterior entropy corresponding to it would increase, making its weight, so its importance to the final recognition, decrease. This is also valid in the case of a complete interruption of one stream. In this case, the entropy should be close to maximum, and the weight assigned to the missing stream close to zero. This practically makes the system revert to one-stream recognition automatically. This process is instantaneous and also reversible, that is, if the missing stream is restored, the entropy would decrease and the weight would increase to the level before the interruption.

9.3 MULTIMODAL SPEAKER LOCALISATION

In the previous section, we have focused on decision-level fusion methods, presenting them in detail because these are the methods that work best for AVSR. In this section, we will shift the focus toward lower-level fusion methods, common in speaker localisation. Just as above, we are presenting practical methods which are quite general and could also be used for other applications.

Speaker localisation consists in finding the active speaker in a video sequence containing several persons. This can be useful especially in videoconference applications and automatic annotation of video sequences. Typically, speaker localisation can be done either in the audio modality using a microphone array or in the visual modality by detecting movement. Multimodal approaches use the correlation between audio and video to find the mouth of the speaker and should be able to make the difference between the active speaker and a person murmuring something inaudible, something that is impossible with a video-only approach. At the same time, the multimodal method should be less affected by noise in the audio, compared to the microphone-array approach.

As a multimodal application, speaker localisation is simpler than speech recognition, in the respect that only correlations are sought between the modalities, without requiring a deeper understanding of the information in the two streams. This is why feature fusion methods work well in this application. The general approach is to transform the data in such a way that a correlation between the audio and a specific location in the video is found, achieving in this way the actual localisation of the sound source. Most of the remaining methods use score-level fusion in an unconventional way, that is, instead of combining scores, probability density models are built on the fly from each modality and then compared using a distance measure, typically mutual information, again looking for areas of high correlation in the video.

From the category of feature-level fusion methods, we can mention the approach of Slaney and Covell [29], who use canonical correlation analysis to find a linear mapping which maximises the audio-visual correlation on training data. Measuring the audio-visual correlation in the transformed space gives a quantitative measure of audio-visual synchrony.

Monaci et al. [30] decompose the image sequence in video atoms and then search for correlations between the audio and the movement of these atoms, to find the speaker. The framework is extended

[31] such that a multimodal dictionary of audio-visual atoms can be learned directly.

As mentioned above, in the category of score-level fusion methods, mutual information is the measure of choice to estimate the level of correlation between the audio and visual modalities. Hershey and Movellan [32] use an estimate of the mutual information between the average acoustic energy and the pixel value, which they assume to be jointly Gaussian.

Nock et al [33, 34] estimate the mutual information between the audio and the video by two methods, one using histograms to estimate the probability densities and the other using multivariate Gaussians.

Fisher et al. [35] use a nonparametric statistical approach to learn maximally informative joint subspaces for multimodal signals. This method requires neither prior model nor training data. In [36], the method is further developed, showing how the audio-visual association problem formulated as a hypothesis test can be related to mutual information-based methods.

Butz and Thiran [37, 38] propose an information theoretic framework for the analysis of multimodal signals. They extract an optimised audio feature as the maximum entropy linear combination of power spectrum coefficients. They show that the image region where the intensity change has the highest mutual information with the audio feature is the speaker's mouth. Besson et al. [39] use the same framework to detect the active speaker among several candidates. The measure that they maximise is the efficiency coefficient, i.e. the ratio between the audio-visual mutual information and the joint entropy. They use optical flow components as visual features, extracting them from candidate regions identified using a face tracker.

Finally, it is possible to use directly score-level fusion, in the case when models of the multimodal signals have been learned on training data [40]. This method uses a trained joint probability density function (pdf) of audio-visual one-dimensional features and then simply finds the location in the video giving the highest score according to this learned model. The advantage is that the amount of computation is greatly reduced at test time, as the computational burden is shifted toward the training phase.

9.4   CONCLUSION

We have shown how the concepts in the previous two chapters can be implemented in practice on two applications, AVSR and multimodal speaker localisation, in both cases leading to an accuracy above that of the corresponding monomodal methods. Most, if not, all the methods presented here are very general because they do not depend on any specific attributes of the modalities used. This makes them applicable to many other problems beside the two presented here.

In the first part, we have focused on high-level fusion methods applied to AVSR. We have also detailed several quality measures that can be used to evaluate the reliability of the audio and visual streams.

In the second part, we have shifted the focus towards lower-level fusion methods, which are more common for audio-visual speaker localisation. Here, we have presented several feature-level fusion methods, along with some score-level fusion methods which use mutual information as a means of combining scores.

The next chapter will present the implementation of one particular fusion method on two applications, the first one being again AVSR, whereas the second is biometric authentication.

REFERENCES

1. G. Potamianos, C. Neti, J. Luettin, I. Matthews, Audio-visual automatic speech recognition: an overview, in: G. Bailly, E. Vatikiotis-Bateson, P. Perrier (Eds.), Issues in Audio-Visual Speech Processing, MIT Press, 2004.
2. L. Rabiner, B. Juang, An introduction to Hidden Markov Models, IEEE ASSP Mag. 3 (1) (1986) 4–16.
3. L. Rabiner, A tutorial on Hidden Markov Models and selected applications in speech recognition, Proc. IEEE 77 (2) (1989), pp. 257–286.
4. R. Bakis, Continuous speech recognition via centisecond acoustic states, in: Proceedings of the 91st Meeting of the Acoustical Society of America, 1976, p. S97.
5. A. Adjoudani, C. Benoît, On the integration of auditory and visual parameters in an HMM-based ASR, in: D. Stork, M. Hennecke (Eds.), Speechreading by Humans and Machines, Springer, 1996, pp. 461–471.
6. S. Young, D. Kershaw, J. Odell, D. Ollason, V. Valtchev, P. Woodland, The HTK Book, Entropic Ltd., Cambridge, 1999.

7. G. Potamianos, H. Graf, Discriminative training of HMM stream exponents for audio-visual speech recognition, in: Proceedings of the International Conference on Acoustics, Speech and Signal Processing, 1998, pp. 3733–3736.

8. S. Nakamura, H. Ito, K. Shikano, Stream weight optimization of speech and lip image sequence for audio-visual speech recognition, Proc. Int. Conf. Spoken Lang. Process. III (2000) 20–23.

9. G. Gravier, S. Axelrod, G. Potamianos, C. Neti, Maximum entropy and MCE based HMM stream weight estimation for audio-visual ASR, in: Proceedings of the International Conference on Acoustics, Speech and Signal Processing (ICASSP), 2002, pp. 853–856.

10. C. Miyajima, K. Tokuda, T. Kitamura, Audio-visual speech recognition using MCE-based HMMs and model-dependent stream weights, Proc. Int. Conf. Spoken Lang. Process. II (2000) 1023–1026.

11. P. Jourlin, Word dependent acoustic-labial weights in HMM-based speech recognition, in: Proceedings of the European Tutorial Workshop on Audio-Visual Speech Processing, 1997, pp. 69–72.

12. E. Sanchez-Soto, A. Potamianos, K. Daoudi, Unsupervised stream weight computation using anti-models, in: Proceedings of the International Conference on Acoustics, Speech and Signal Processing, 2007.

13. S. Tamura, K. Iwano, S. Furui, A stream-weight optimization method for audio-visual speech recognition using multi-stream HMMs, Proc. Int. Conf. on Acoust. Speech Signal Process. 1 (2004) 857–860.

14. S. Tamura, K. Iwano, S. Furui, A stream-weight optimization method for multi-stream HMMs based on likelihood value normalization, in: Proceedings of the International Conference on Acoustics, Speech and Signal Processing, 2005, pp. 468–472.

15. N. Fox, R. Gross, J. Cohn, R. Reilly, Robust biometric person identification using automatic classifier fusion of speech, mouth, and face experts, IEEE Trans. Multimedia 9 (4) (2007) 701–714.

16. S. Cox, I. Matthews, A. Bangham, Combining noise compensation with visual information in speech recognition, in: Proceedings of the European Tutorial Workshop on Audio-Visual Speech Processing, 1997, pp. 53–56.

17. S. Gurbuz, Z. Tufekci, E. Patterson, J. Gowdy, Application of affine-invariant Fourier descriptors to lip-reading for audio-visual speech recognition, in: Proceedings of the International Conference on Acoustics, Speech and Signal Process. 2001, pp. 177–180.

18. P. Teissier, J. Robert-Ribes, J. Schwartz, Comparing models for audiovisual fusion in a noisy-vowel recognition task, IEEE Trans. Speech Audio Process. 7 (1999) 629–642.

19. M. Heckmann, F. Berthommier, K. Kroschel, Noise adaptive stream weighting in audio-visual speech recognition, EURASIP J. Appl. Signal Process. 2002 (2002) 1260–1273.

20. U. Meier, W. Hurst, P. Duchnowski, Adaptive bimodal sensor fusion for automatic speechreading, in: Proceedings of the International Conference on Acoustics, Speech and Signal Processing, 1996, pp. 833–836.

21. S. Dupont, J. Luettin, Audio-visual speech modelling for continuous speech recognition, IEEE Trans. Multimedia 2 (2000) 141–151.

22. F. Berthommier, H. Glotin, A new SNR-feature mapping for robust multistream speech recognition, in: Proceedings of the International Congress on Phonetic Sciences, 1999, pp. 711–715.

23. H. Glotin, D. Vergyri, C. Neti, G. Potamianos, J. Luettin, Weighting schemes for audio-visual fusion in speech recognition, in: Proceedings of the International Conference on Acoustics, Speech and Signal Processing, 2001, pp. 173–176.

24. G. Potamianos, C. Neti, Stream confidence estimation for audio-visual speech recognition, in: Proceedings of the International Conference on Spoken Language Processing (ICSLP), 2000, pp. 746–749.

25. G. Potamianos, C. Neti, G. Gravier, A. Garg, A. Senior, Recent advances in the automatic recognition of audio-visual speech, Proc. IEEE 91 (9) (2003), pp. 1306–1326.

26. R. Seymour, D. Stewart, J. Ming, Audio-visual integration for robust speech recognition using maximum weighted stream posteriors, in: Proceedings of Interspeech, 2007, pp. 654–657.

27. H. Misra, H. Bourlard, V. Tyagi, New entropy-based combination rules in HMM/ANN multi-stream ASR, in: Proceedings of the IEEE International Conference on Acoustics, Speech, and Signal Processing (ICASSP), 2003, pp. 741–744.

28. M. Gurban, J. Thiran, T. Drugman, T. Dutoit, Dynamic modality weighting for multi-stream HMMs in Audio-Visual Speech Recognition, in: Proceedings of the 10th International Conference on Multimodal Interfaces, 2008.

29. M. Slaney, M. Covell, Facesync: a linear operator for measuring synchronization of video facial images and audio tracks, Neural Inf. Process. Syst., Volume 13 (2001) 814–820.

30. G. Monaci, O.D. Escoda, P. Vandergheynst, analysis of multimodal sequences using geometric video representations, Signal Process. 86 (12) (2006) 3534–3548.

31. G. Monaci, P. Jost, P. Vandergheynst, B. Mailhe, S. Lesage, R. Gribonval, Learning multi-modal dictionaries, IEEE Trans. Image Process. 16 (9) (2007) 2272–2283.

32. J. Hershey, J. Movellan, Audio vision: using audio-visual synchrony to locate sounds, Neural Inf. Process. Syst., Volume 12 (2000) 813–819.

33. H. Nock, G. Iyengar, C. Neti, Assessing face and speech consistency for monologue detection in video, in: Proceedings of ACM Multimedia, 2002, pp. 303–306.

34. H. Nock, G. Iyengar, C. Neti, Speaker localisation using audio-visual synchrony: an empirical study, in: Proceedings of the International Conference on Image and Video Retrieval, 2003, pp. 488–499.

35. J. Fisher III, T. Darrell, W. Freeman, P. Viola, Learning joint statistical models for audio-visual fusion and segregation, Adv. Neural Inf. Process. Syst. 2001.

36. J. Fisher III, T. Darrell, Speaker association with signal-level audiovisual fusion, IEEE Trans. Multimedia 6 (3) (2004) 406–413.

37. T. Butz, J. Thiran, Feature space mutual information in speech-video sequences, Proc. IEEE Int. Conf. Multimedia Expo 2 (2002) 361–364.

38. T. Butz, J. Thiran, From error probability to information theoretic (multi-modal) signal processing, Signal Process. 85 (2005) 875–902.

39. P. Besson, M. Kunt, T. Butz, J. Thiran, A multimodal approach to extract optimized audio features for speaker detection, in: Proceedings of the European Signal Processing Conference (EUSIPCO), 2005.

40. M. Gurban, J. Thiran, Multimodal speaker localization in a probabilistic framework, in: Proceedings of the 14th European Signal Processing Conference (EUSIPCO), 2006.

Chapter 10

A Multimodal Recognition Framework for Joint Modality Compensation and Fusion

Konstantinos Moustakas, Savvas Argyropoulos and
Dimitrios Tzovaras
Informatics and Telematics Institute/Centre for Research and
Technology Hellas, Thessaloniki, Greece

Multimodal Signal Processing, ISBN: 9780123748256

In Chapter 7, the methods for multimodal integration have been categorised and presented in detail. In Chapter 8, these theoretical methods have been illustrated on two examples. In this chapter, we will go further into practical applications and show two instances of how a method of multimodal integration can be implemented, this time aiming to concretise the concepts previously presented. The first example shown is that of enhanced speech recognition, with a particular focus on the disabled, while the second is biometric authentication. Both applications use the same joint modality recognition framework that is detailed in the following. The purpose here is to show how, from the multitude of fusion methods presented earlier, one can be chosen and implemented for two quite different problems.

10.1   INTRODUCTION

The widespread deployment of novel human–computer interaction methods has changed the way individuals communicate with computers. Since Sutherland's SketchPad in 1961 or Xerox' Alto in 1973, computer users have long been acquainted with more than the traditional keyboard to interact with a system. More recently with the desire of increased productivity, of seamless interaction and immersion, of e-inclusion of people with disabilities, as well as with the progress in fields such as multimedia/multimodal signal analysis and human–computer interaction, multimodal interaction has emerged as a very active field of research (e.g., [1, 2]).

Multimodal interfaces are those encompassing more than the traditional keyboard and mouse. Natural input modes are put to use (e.g., [3, 4]), such as voice, gestures and body movement, haptic interaction, facial expressions and more recently physiological signals. As described in [5] multimodal interfaces should follow several guiding principles: multiple modalities that operate in different spaces need to share a common interaction space and to be synchronised; multimodal interaction should be predictable and not unnecessarily complex and should degrade gracefully for instance by providing for modality switching; finally multimodal interfaces should adapt to user's needs, abilities, and environment.

A key aspect in many multimodal interfaces is the integration of information from several different modalities to extract high-level information non-verbally conveyed by users or to identify cross-correlation between the different modalities. This information can then be used to transform one modality to another. The potential benefits of such modality transformations to applications for disabled users are very important because the disabled user may now perceive information that was not previously available, using an alternate communication channel. This implies that, e.g., blind users may become able to communicate with hearing impaired users, as illustrated in Figure 10.1.

Another very interesting research perspective is the multimodal signal processing so as to improve the recognition rate of multimodal

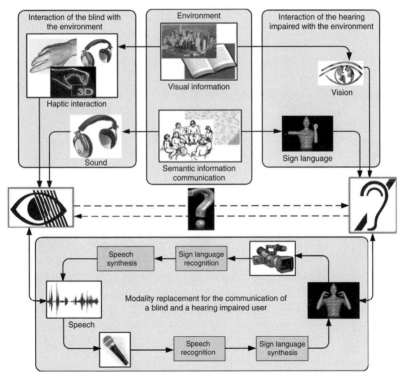

FIGURE 10.1   Architecture of a cross-modal transformation system for the blind and the hearing impaired.

actions by jointly recognising the user action in both communication channels exploiting their mutual information. An example of such a system could be the audio-visual speech recognition that is also described later in the chapter, where speech is recognised both in the audio channel and the visual channel using the information obtained from the lip shape. Another approach could be the joint modality recognition of sign language broadcasting, where the signs are visually recognised along with the displayed subtitles, that could increase recognition rate and could also solve synchronisation problems. These approaches apart from theoretically increasing the recognition rate are very useful for applications for the disabled, because they usually cannot use all communication channels or cannot perform very well thus inhibiting the recognition rates, by using the complementary modalities to replace or to compensate possible errors in the problematic communication channel.

10.2   JOINT MODALITY RECOGNITION AND APPLICATIONS

Joint modality recognition refers to the identification of specific patterns using as input two or more modalities. Typical examples include the recognition of spoken language using the speech and the lip shape modality, the recognition of pointing direction using gaze and gestures, etc. Joint modality recognition can be useful in a variety of applications and is of particular interest for applications for the disabled. The basic concept of most human–computer interaction applications that focus on the disabled is the idea of modality compensation and replacement, which is the use of information originating from various modalities to improve or to compensate for the missing input modality of the system or the users as illustrated in Figure 10.2.

Figure 10.1 illustrates the architecture of a prospective framework for computer-aided interaction of blind and hearing-impaired users with the environment as well as with each other. Visual information about the environment has to be conveyed to the blind user via the haptic and/or the auditory channel, while communication and the acquisition of various semantic information can be performed

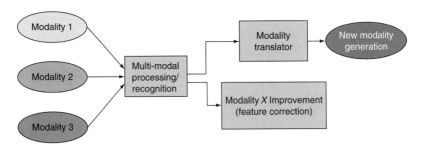

FIGURE 10.2   Flow chart of the application.

using natural language. On the other hand, the hearing-impaired user acquires visual information using vision and communicates with other people using sign language.

It is obvious that communication between a blind and a hearing-impaired user is not possible using physical means. Ideally, as illustrated in Figure 10.1, a modality replacement system could be used to recognise all spoken language input of the blind user, convert it into sign language and present it to the hearing-impaired user through an animated avatar. Similarly, sign language gestures should be recognised and converted into text that should then be synthesised into speech using text-to-speech synthesis techniques (see Chapter 3).

Such systems for the disabled could significantly benefit from joint modality recognition. The basic idea is to exploit the correlation between modalities to enhance the perceivable information by an impaired individual who cannot perceive all incoming modalities. In that sense, a modality, which would not be perceived due to a specific disability, can be employed to improve the information that is conveyed in the perceivable modalities and increase the accuracy rates of recognition. The results obtained by jointly fusing all the modalities outperform those obtained using only the perceived modalities.

The basic architecture of a modality compensation and replacement approach is depicted in Figure 10.2. The performance of such a system is directly dependent on the efficient multimodal processing of two or more modalities and the effective exploitation of their complementary nature and their mutual information to achieve accurate recognition of the transmitted content. After the recognition has been performed effectively, either a modality translator can be employed to generate a new modality or the output can be used to detect and correct possibly erroneous feature vectors that may correspond to different modalities. The latter could be very useful in self-tutoring applications. For example, if an individual practices sign language, the automatic recognition algorithm could detect incorrect hand shapes (based on audio and visual information) and indicate them so that the user can identify the wrong gestures and practice more on them. Another example is the joint modality recognition of information like speech, where both the speech and the lip shape should be consistent with each other and a possible misclassification of the one modality could be compensated by the other one. Similar processing can be also applied to recognition of sign language [6] by the joint modality recognition of signs and facial expressions.

In the literature there exist several interesting approaches that exploit modality replacement to develop aids for the disabled that could benefit from a joint modality recognition scheme. In [7], a system has been developed that allows the communication between blind and hearing-impaired users in the context of a treasure hunting game that is played collaboratively by the users, thus partially implementing the framework of Figure 10.1.

In [8], a haptic browser has been developed that allows the navigation of blind users in the World Wide Web by converting the information that is presented on the websites to haptic representations in a virtual 3D space. The user can access the specific components of the website through a haptic device, whereas semantic information is rendered using text-to-speech synthesis. Other interesting approaches included real-time haptic [9] and aural [10] navigation in unknown environments.

The framework described in the following focusses on a general joint modality recognition system using a novel efficient scheme that is based on channel coding. In Section 10.3, some fundamental theoretical issues are presented, while Section 10.4 describes the application of the presented framework in audio-visual speech recognition. Section 10.5 demonstrates the wide applicability of the framework by briefly describing a system for joint face-gait recognition in a biometric system.

10.3 A NEW JOINT MODALITY RECOGNITION SCHEME

10.3.1 Concept

Typical unimodal pattern recognition approaches that can be found in the literature adopt the following procedure (see Chapter 2): Initially, after segmenting the area of interest in an image, in a video or more generally in a signal, features are extracted that represent as much as possible the pattern to be recognised. These initial features (signature features) are stored in a database for future reference. When processing a signal during run-time the same features are extracted and compared to the initial signatures so as to identify possible similarities.

For the case of joint modality recognition, more than one feature are available and they have to be carefully combined so as to exploit their interdependencies and/or complementarity. Chapter 7 presents in detail the multimodal fusion methods that can be used to achieve this.

The fusion mechanism presented here is in between feature-level and score-level fusion, both detailed in Chapters 7 and 8. This approach combines the characteristics of the features of all the modalities so as to provide a single unified recognition result. The algorithm is based on channel coding and is analysed in the following.

10.3.2 Theoretical Background

For the rest of this chapter, capital symbols will denote stochastic sequences and small symbols will denote their respective realisations.

The Slepian–Wolf theorem [11] addresses the problem of coding distributed (not co-located) sources and decoding them jointly, as depicted in Figure 10.3a. If we consider two random sequences X and Y that are encoded using separate conventional entropy encoders and decoders, the achievable rates are $R_X \geq H(X)$ and $R_Y \geq H(Y)$, where $H(X)$ and $H(Y)$ are the entropies of X and Y, respectively. However, if the two sequences are jointly decoded the achievable rate region according to the Slepian–Wolf theorem is defined by [11]:

$$R_X \geq H(X|Y), \quad R_Y \geq H(Y|X), \quad R_X + R_Y \geq H(X,Y) \quad (10.1)$$

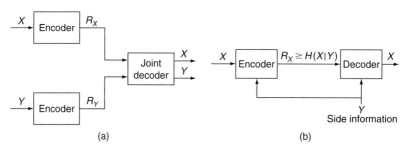

FIGURE 10.3   Encoder–decoder.

where $H(X|Y)$ and $H(Y|X)$ are the conditional entropies and $H(X,Y)$ is the joint entropy of X and Y.

The Slepian–Wolf theorem can be also applied in the problem of source coding with decoder side information (Figure 10.3b). Specifically, if the sequence X is correlated with the sequence Y, which is available only at the decoder, but not at the encoder, the achievable rate for sequence X is $R_X \geq H(X|Y)$. Thus, even though the encoder does not have access to the correlated sequence Y, it can compress source X as if Y were available at the encoder. However, the Slepian–Wolf theorem does not provide a practical implementation of the described system.

Pattern recognition can be formulated as a problem of source coding with decoder side information if we consider the stored signature (gallery) of a pattern and the real-time acquired signature (probe) as the random variables X and Y, respectively. This representation is reasonable because these signals are correlated and the Y is only available at the decoder (recognition) side. Let x^1 be the original representation of the pattern b at the signature extraction stage at time t. In general, the probe and gallery data are not identical even in the case of dealing with the same pattern due to time-related modifications in the pattern, its presentation and the sensor that captures the raw data. The noise induced in the signal b' can be modelled by a (virtual) additive noise (or correlation) channel that induces noise w. Thus, at the recognition stage, which takes place at time t', the recognition system needs to detect whether the input signal $y = x + w$ refers to a specific pattern or not.

This model is analogous to data communication over noisy channels and is similar to the notion that Slepian–Wolf coding protects X for 'transmission' over the (virtual) noisy channel. At the decoder, Y is regarded as if it were X after transmission over the noisy channel and corrects it using error correcting codes. Intuitively, the noise w_g induced by the channel in case of a correct pattern is small, whereas the noise w_i in other patterns is relatively large. Thus, the channel decoder can decode the codeword only when the induced noise is small and the error is within the correcting capabilities of the channel code. Otherwise, if the noise of the channel corrupts the signal, the resulting codeword cannot be decoded and the pattern is considered as non-matching. If the selected error correcting code is suitable for error protection on this channel, the decoder will decode X errorlessly and the pattern will be recognised.

In this work, we extend the Slepian–Wolf theorem to the case of four correlated sources X_1, X_2, X_3 and X_4 to handle multimodal signals. Let R_i denote the rate for X_i, $i \in \{1, 2, 3, 4\}$, then from the extension of the Slepian–Wolf theorem to multiple sources the achievable rate region is

$$R(\mathbf{S}) > H\big(X(\mathbf{S}) \,|\, X(\mathbf{S}^C)\big) \tag{10.2}$$

where $\mathbf{S} \subseteq \{1, 2, 3, 4\}$, $R(\mathbf{S}) = \sum_{l \in \mathbf{S}}$ and $X(\mathbf{S}) = \{X_l : l \in \{1, 2, 3, 4\}\}$.

Figure 10.4 illustrates the architecture of the joint modality recognition method. At the signature extraction (encoding stage), the sequences X_1, X_2, which represent the feature vectors of the two signals, are encoded using two separate Slepian–Wolf encoders. The signatures (or templates), which are stored in the database of the system, consist of the generated codewords S_1, S_2. The recognition stage (decoding) consists of two steps: first, X_1 is decoded using the corresponding decoder and X_3, which is the signature corresponding to X_1 provided during the recognition stage. Then, the estimated version $\hat{X}_1$ of X_1 together with X_4 are the input to the second decoder to estimate X_2.

A critical parameter in the design of the system is the rate of the encoders. On one hand, a high rate generates long codewords that is expected to decrease the false acceptance ratio (FAR) but to also increase the false rejection ratio (FRR). On the other hand, a low rate

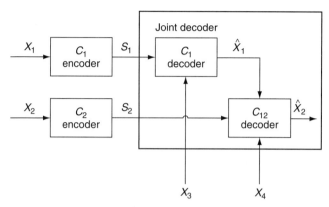

FIGURE 10.4   Joint modality recognition.

generates small codewords, which is expected to increase the FAR but to decrease the FRR. Thus, the design of an effective recognition system based on channel codes involves the careful selection of the channel code rate to achieve the optimal performance.

The minimum rate for the representation of the encoded sequences is given by equation (10.2). Since X_3 and X_4 are available at the decoder, X_1 and X_2 can be compressed at rates R_1 and R_2 lower bounded by equation (10.2). Thus, the rate R_1 of the first encoder is lower bounded by $R_1 \geq H(X_1|X_2)$ and the rate R_2 of the second encoder is bounded by $R_2 \geq H(X_2|X_1, X_3, X_4)$.

10.4 JOINT MODALITY AUDIO-VISUAL SPEECH RECOGNITION

Audio-visual speech recognition could be a very significant aid for the hearing impaired. Even if synchronised audio and video content is the main form of digital communication, hearing-impaired users cannot perceive the audio information. However, as described in Section 10.2, a joint modality recogniser of speech and lip shape could be used so as to use the non-perceivable information (i.e., speech) to enhance the information that can be accessed by the hearing-impaired user (i.e., video, lip shape) and provide as output information about the spoken dialogues in text. Moreover, such a system could be further used to resolve the ambiguities raised by the lip shapes according to the context of the spoken dialogues.

The architecture of the joint modality audio-visual speech recognition system is depicted in Figure 10.5. At the signature extraction stage, the lip shape and speech feature vectors X_1 and X_2 are initially extracted using LDA and MFCC/HMMs, respectively. For a detailed description on the extraction of the features the reader is referred to [12]. The extracted feature vectors are encoded using a channel encoder. It must be stressed that the rate of the LDPC encoders in Figure 10.5 is different for each modality according to equation (10.2). The resulting codewords S_1 and S_2 comprise the signatures of the modalities and are stored in the database of the system.

In the recognition stage, the lip shape and speech feature vectors X_3 and X_4 are extracted. Subsequently, the syndromes S_1 and S_2 that correspond to the claimed identity are retrieved from the database and are fed to the LDPC decoders. These processes are analytically described later.

It must be noted that two unimodal systems based on this scheme could be used to encode the signatures independently. However,

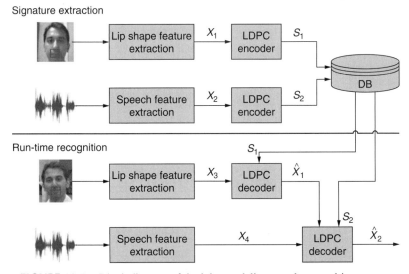

FIGURE 10.5   Block diagram of the joint modality speech recognition system.

the main difference between this scheme and the unimodal case is that the extension of Slepian–Wolf theorem to multiple sources [equation (10.2)] is employed, which limits the rate required to represent the templates. Otherwise, if the unimodal recognition scheme had been used for every modality independently, the rate required to code each feature vector would be higher. This in turn would affect the size of the signatures and the performance of the system.

10.4.1 Signature Extraction Stage

Initially, at the signature extraction stage, the signatures of an individual concerning the lip shape and speech modalities for each pattern are obtained. The extracted features form the vector $x_i = \left[x_1^i, \ldots, x_{k_i}^i \right]$, $i \in \{1, 2\}$ thus $x_i \in \mathfrak{R}^{k_i}$. The feature vector x_i must be transformed from the continuous to the discrete domain so that it can be further processed by the channel encoder. This mapping can be represented by a uniform quantiser with 2^{L_i} levels. Each component of x_i is then mapped to an index in the set Q, through the function $u : \mathfrak{R}^{k_i} \to Q^{k_i}$, where $Q = \{0, 1, \ldots, L_i - 1\}$. Each one of the resulting vectors $q_i = u(x_i)$ is fed to the Slepian–Wolf encoder, which performs the mapping $e : Q^{k_i} \to C^{n_i}$, where $C = \{0, 1\}$ and outputs the codeword $c_i = e(q_i)$, $c_i \in C^{n_i}$.

In this work, the Slepian–Wolf encoder is implemented by a systematic LDPC encoder [13]. LDPC codes were selected because of their excellent error detecting and correcting capabilities. They also provide near-capacity performance over a large range of channels, while simultaneously admitting implementable decoders. An LDPC code (n, k) is a linear block code of codeword length n and information block length k, which is defined by a sparse $(n - k) \times n$ parity matrix H, where $n - k$ denotes the parity bits produced by the encoder. The code rate is defined as $r = k/n$. A code is a systematic code if every codeword consists of the original k-bit information vector followed by $n - k$ parity-bits. In the system presented here, the joint bit-plane encoding scheme of [14] was employed to avoid encoding and storing the L_i bit-planes of the vector q_i separately. Alternatively, LDPC codes in a high-order Galois-field could be employed, but binary LDPC codes (GF(2)) were selected due to ease of implementation.

Subsequently, the k_i systematic bits of the codeword $\mathbf{c}_i$ are discarded and only the *syndrome* $\mathbf{s}_i$, that is the $n_i - k_i$ parity bits of the codeword $\mathbf{c}_i$, is stored to the database. Thus, the signatures consist of the syndromes $s_i = [c_{k_i+1} \ldots c_{n_i}]$, $\mathbf{s}_i \in C^{(n_i-k_i)}$, and their size is $(n_i - k_i)$. It must be stressed that the rate of the two LDPC encoders is different because the statistical properties of the two modalities are different.

10.4.2   Recognition Stage

At the recognition stage, a new signature is extracted from the features, and the vector $x_i = \left[x_1^i, \ldots, x_{k_i}^i \right]$, $i \in \{3, 4\}$, $x_i \in \mathfrak{R}^{k_i}$, is constructed. The vectors x_3 and x_4, which form the side information corresponding to x_1 and x_2, respectively, are fed to the LDPC decoder. The decoding function $d : C^{(n_i-k_i)} \times \mathfrak{R}^{k_i} \to Q^{k_i}$ combines x_i, $i \in \{3, 4\}$ with the corresponding syndromes which are retrieved from the database and correspond to the pattern I. The decoder employs belief-propagation [15, 16] to decode the received codewords.

If the errors introduced in the side information with regard to the originally encoded signal are within the error correcting capabilities of the channel decoder, then the correct codeword is output after a number of iterations and the specific pattern is recognised. More specifically, the lip shape feature vector x_3 is initially fed to the LDPC decoder. The output of the LDPC decoder is the quantised vector $\hat{q}_1 = d(s_1, x_3)$. In general, besides the code rate, the error correcting capabilities of the channel decoder also depend on the information of the noisy channel and the relationship between the noise induced by the channel and the side information. Accurate modelling of the distribution of the noisy channel can improve the knowledge of the channel decoder by exploiting a priori information.

Subsequently, the decoded codeword $\mathbf{q}_1$ is fed to the speech LDPC decoder. Then, the decoding function combines x_4, $\hat{x}_1$ and s_2 to decode the original codeword x_2, thus $\hat{q}_2 = d(x_4, s_2, \hat{x}_1)$. The correlation between x_1 and x_2 can be modelled by a binary symmetric channel (BSC) with crossover probability p, which is unique for each user action and is stored in the database. To detect whether a codeword is correctly decoded, we add 16 cyclic redundancy check (CRC) bits at the beginning of the feature vector. By examining these bits the integrity of the original data is detected. If the codeword is correctly decoded, then the specific pattern is recognised. Otherwise, if the decoder cannot decode the codeword (which is indicated if the number of iterations increases over a specific number N_{iter}), the input pattern is a non-match.

Speech recognition rates are seen to be significantly improved using the presented framework. Even if an extensive evaluation study has not been performed, there has been an increment in the recognition rate of both speech and speech commands over 3 and 5%, respectively.

10.5   JOINT MODALITY RECOGNITION IN BIOMETRICS

In the previous section, we have shown how modality integration can be implemented through the suggested joint recognition framework in the case of audio-visual speech recognition. To prove the wide applicability of this scheme, we present in the following another example, a joint modality biometric authentication system.

10.5.1   Overview

The architecture of the multimodal biometric authentication system is depicted in Figure 10.6.

At the enrolment stage, the face and gait feature vectors X_1 and X_2 are initially extracted as described in [17, 18], respectively. The extracted feature vectors are encoded using a channel encoder. The resulting codewords (syndromes) S_1 and S_2 comprise the biometric templates of the modalities and are stored in the database of the system. Thus, if the database of the biometric system is attacked, the attacker cannot access the original raw biometric data or their corresponding features but only S_1 and S_2, which cannot reveal any information as it will be explained in the following.

At the authentication stage, the face and gait feature vectors X_3 and X_4 are extracted. Subsequently, the syndromes S_1 and S_2 that correspond to the claimed identity are retrieved from the database and are fed to the LDPC decoders.

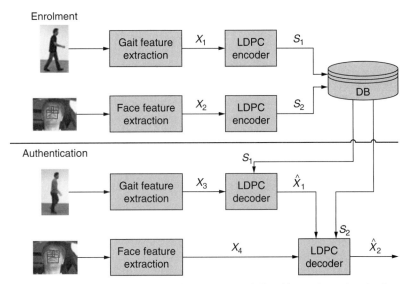

FIGURE 10.6 Block diagram of the joint modality biometric authentication system.

The enrolment and the authentication procedures are implemented exactly as the signature extraction and run-time recognition described in Sections 10.4.1 and 10.4.2, respectively, by substituting the feature vectors of the lip shape and the speech modalities with the gait and the face modalities.

10.5.2 Results

Because of the availability of face and gait databases, an extensive evaluation study has been performed for this application scenario. The multimodal database was created by aggregating two unimodal databases for face and gait, as suggested in [19].

The test database contains 29 different individuals recorded under five different conditions giving 145 sequences in total. Although the number of subjects might seem small at first, one must consider that these individuals were recorded under varying conditions such as different facial expressions and different illumination conditions. As whole image sequences are available for each individual, multiple feature vectors are extracted from one sequence. This number varies between 50 and 300 depending on the number of successful extractions per sequence. From the whole corpus a subset was selected to prove the effectiveness of the fusion approach. While matching feature vectors extracted within the same sequence produces low error rates in general, inter-sequence matches are a more challenging task especially when the variances present in the enrolment sequence differ much from these present in the matching sequence. A moderately challenging combination of enrolment and matching conditions was chosen for the evaluation: subjects with neutral facial expression were matched against talking subjects.

The gait database was captured in an indoor environment and consists of 75 people walking in a predefined path in a front-parallel view from the camera. The main course of walking is around 6 m and the distance from the camera varies from 4 to 6 m. In addition, for each sequence, the 3D depth map was captured using a stereo camera. This is the first database that has depth data for assisted gait recognition. For each subject, two different conditions were captured: (a) the 'normal' condition, and (b) the 'hat' condition in which the users wear a hat (e.g., there is a slight change in appearance apart from different clothing). The 'normal' set was used as the gallery set and the other set was considered as the probe set.

For the creation of the multimodal database, the maximum number of virtual subjects is determined by the size of the smallest unimodal database, thus its population is 29 subjects. The virtual subjects were obtained with natural ordering within each unimodal database. In other words, the N-th virtual user was created using the N-th user trait from each database. Thus, the multimodal database consists of 29 subjects and two recordings. The evaluation was performed using the first half of the subjects (all recordings) for training and the other half of the subjects for testing. Thus, the test set contains subjects that have not been used for training. The sets slide for each run by one subject and in that way the training-testing dataset combinations that are created are equal to the number of subjects. In particular, for each run, 15 subjects were used for training and the remaining 14 were used for testing. Thus, the total number of genuine and impostor transactions in the training set is $15 \times 29 = 435$ and $15 \times 14 \times 29 = 6090$, respectively. The test set contains $14 \times 29 = 406$ genuine and $14 \times 13 \times 29 = 5278$ impostor transactions.

In an authentication scenario (or verification), the biometric system is used to grant access to individuals. Initially a subject claims his/her identity and the gait system compares the signature with the stored one in the database. Then, based on the authentication procedure, the system establishes whether the identity of the user is the claimed one. In this respect, authentication results in a one-to-one comparison and is quite different from the identification scenario, in which the system has to determine the identity of users by comparing the measured data with all the enrolled data in the database (one-to-many database). The performance of the biometric system is evaluated in terms of the false acceptance rate (FAR), the false rejection rate (FRR) and the equal error rate (EER), which corresponds to the point where the FAR is equal to FRR. FAR measures the ratio of impostors who are incorrectly accepted into the system as legitimate users and FRR measures the ratio of genuine users who are rejected as impostors. Also, results are presented using rate operating characteristic (ROC) curves, which present the verification rate (or genuine acceptance rate, GAR) versus the FAR. The FRR can then be computed as 1-GAR.

Figure 10.7a reports the performance results of the gait authentication system as a function of the security bits using the proposed scheme for the protection of the templates. Thus, the horizontal axis represents the numbers of the syndrome bits, while the vertical axis represents the FAR and FRR. The more bits used for the syndrome, the more secure is the template because it is more difficult to be broken. On the other hand, increasing the size of the syndrome increases the sensitivity of the system, which results in more authentication failures of legitimate users. Thus, the recognition accuracy of the proposed system can be determined by specifying the code rate r. This is similar to the conventional approach that determines the operating points of the ROC curve by varying the threshold that determines which subjects are granted access. The reported results are also compared with the method presented in [18] using ROC curves, as depicted in Figure 10.7b. It can be seen that the proposed scheme achieves slightly better performance while at the same time it provides security to the stored templates.

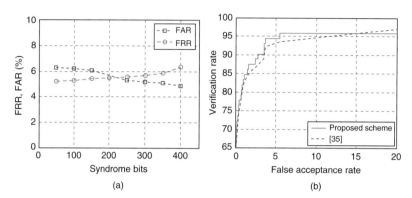

FIGURE 10.7 (a) FAR and FRR as a function of the syndrome bits and (b) ROC curve of the gait authentication system.

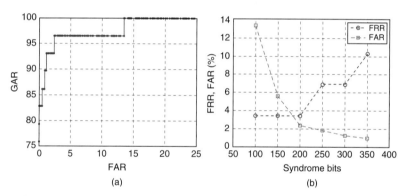

FIGURE 10.8 (a) ROC curve of the face authentication system and (b) FAR and FRR as a function of the syndrome bits.

Furthermore, the performance of the face authentication system is illustrated in Figure 10.8. Specifically, Figure 10.8a shows the ROC curve of the face module. It must be stressed that if the authentication was based on estimating the Euclidean distances between the gallery and probe feature vectors (rather than using the proposed methodology), the performance of the face classifier would be exactly the same. Thus, it is obvious that the proposed scheme provided template security at no cost in the performance of the face classifier. Moreover, Figure 10.8b depicts the FAR and FRR rates as a function of the size of the face template. Similar to the gait module, as the size of the templates increases the FRR increases and FAR decreases.

Although matching feature vectors extracted within the same sequence produces low error rates in general, inter-sequence matches are a more challenging task especially when the variances present in the enrolment sequence differ much from these present in the matching sequence. Face recognition algorithms are known to be sensitive to data outliers in terms of incorrectly normalised samples and severe intrapersonal variations such as extreme facial expressions and occlusions caused by e.g., glasses and beards. Algorithms are only robust to a certain degree of intrapersonal variations. It is a difficult problem to guarantee proper input to a recognition algorithm, quality measures, e.g., ensuring proper localisation of the eyes have proven to be a feasible approach to ensure proper input for autonomous recognition systems [20]. A moderately challenging combination of enrolment and matching conditions was chosen for the evaluation: subjects with neutral facial expression were matched against talking subjects.

It must be also noted that throughout all the experiments, the same global thresholds and set of parameters were used. Thus, we may conclude that the performance of the system will not change if more users are enrolled or removed from the system.

10.6 CONCLUSIONS

Multimodal human–computer interaction has been a very active field of research during the latest years. Processing multimodal signals instead of unimodal can offer many advantages to a large set of applications including aids for the disabled that vary from modality replacement aids, where information originating from various modalities not perceivable by a disabled user are transformed into a modality that is perceivable, to joint modality recognition systems that use more than one correlated modalities so as to achieve higher recognition rates and robustness. This chapter presented two concrete applications of multimodal fusion, both using a joint modality scheme using ideas from the area of channel coding. The approach could be used for a variety of multimodal pattern recognition applications, where the input sources exhibit slight, or in some cases no, correlation.

REFERENCES

1. "w3c" workshop on multimodal interaction, Sophia Antipolis, France. <http://www.w3.org/2004/02/mmi-workshop-cfp.html>, 2004 (July 2009).
2. J. Strickon, Guest (Ed.), Interacting with emerging technologies, IEEE Comput. Graph. Applic. Jan-Feb 2004, pp. 1–2 (special issue).
3. I. Marsic, A. Medl, J. Flanagan, Natural communication with information systems, in: Proc. IEEE. 88 (8) (2000) 1354–1366.
4. J. Lumsden, S.A. Brewster, A paradigm shift: Alternative interaction techniques for use with mobile & wearable devices, in: Proceedings of 13th Annual IBM Centers for Advanced Studies Conference CASCON'2003, Toronto, Canada, 2003, pp. 97–110.
5. T.V. Raman, Multimodal interaction design principles for multimodal interaction, in: CHI 2003, Fort Lauderdale, USA, 2003, pp. 5–10.
6. J. Lichtenauer, E. Hendriks, M. Reinders, Sign language recognition by combining statistical dtw and independent classification, IEEE Trans. Pattern Anal. Mach. Intell. 30 (11) (2008) 2040–2046.
7. K. Moustakas, D. Tzovaras, L. Dybkjaer, N. Bernsen, A modality replacement framework for the communication between blind and hearing impaired people, in: Proceedings of the HCI International, San Diego, July 2009, pp. 226–235.
8. N. Kaklanis, D. Tzovaras, K. Moustakas, Haptic navigation in the world wide web, in: HCI International 2009, San Diego, July 2009, pp. 707–715.
9. D. Dakopoulos, S. Boddhu, N. Bourbakis, A 2D vibration array as an assistive device for visually impaired, in: Proceedings of the 7th IEEE International Conference on Bioinformatics and Bioengineering, October 2007, pp. 930–937.
10. T. Pun, G. Bologna, B. Deville, M. Vinckenbosch, Transforming 3D coloured pixels into musical instrument notes for vision substitution applications, EURASIP J. Image Video Process. Spec. Issue Image Video Process. Disabil. 24 (2007) 14–27.
11. D. Slepian, J. Wolf, Noiseless coding of correlated information sources, IEEE Trans. Inf. Theory. 19 (4) (1973) 471–480.
12. A. Nefian, L. Liang, X. Pi, X. Liu, K. Murphy, Dynamic Bayesian networks for audio-visual speech recognition, EURASIP J. Appl. Signal Process. 2002 (11) (2002) 1274–1288.
13. R. Gallager, Low-Density Parity-Check Codes. MIT press, Cambridge, MA, 1963.
14. D. Varodayan, A. Mavlankar, M. Flierl, B. Girod, Distributed grayscale stereo image coding with unsupervised learning of disparity, in: Proceedings of Data Compression Conference, Snowbird, UT, 2007, pp. 143–152.
15. W. Ryan, An introduction to LDPC codes, in: B. Vasic, E.M. Kurtas (Eds.), CRC Handbook for Coding and Signal Processing for Recording Systems, Boca Raton, FL, CRC, 2005.

16. J. Yedidia, W. Freeman, Y. Weiss, Generalized belief propagation, in: Advances in Neural Information Processing Systems, Denver, CO, 2000, pp. 689–695.
17. B. Moghaddam, T. Jebara, A. Pentland, Bayesian face recognition, Pattern Recognit. 33 (11) (2000) 1771–1782.
18. D. Ioannidis, D. Tzovaras, I.G. Damousis, S. Argyropoulos, K. Moustakas, Gait recognition using compact feature extraction transforms and depth information, IEEE Trans. Inf. Forensics Secur. 2 (3) (2007) 623–630.
19. N. Poh, S. Bengio, Using chimeric users to construct fusion classifiers in biometric authentication tasks: an investigation, in: Proceedings of IEEE International Conference on Acoustic, Speech and Signal Processing, Toulouse, France, May 2006, pp. 1077–1080.
20. S. Li, Face detection, in: J.A.K. Li, Stan Z. (Eds.), Handbook of Face Recognition, Springer-Verlag, Berlin, Germany, 2004, pp. 13–38.

Chapter 11

Managing Multimodal Data, Metadata and Annotations: Challenges and Solutions

Andrei Popescu-Belis
Idiap Research Institute, Martigny, Switzerland

Multimodal Signal Processing, ISBN: 9780123748256

11.1 INTRODUCTION

The application of statistical learning to multimodal signal processing requires a significant amount of data for development, training
and test. The availability of data often conditions the possibility
of an investigation and can influence its objectives and application
domains. The development of innovative multimodal signal processing methods depends not only on raw data recorded from various
sources, but more specifically on high added-value data that is accompanied by ground-truth metadata or annotations. This information,
generally added or validated manually, is used for supervised learning
and for testing.

The goal of this chapter is to outline the main stages in multimodal
data management, starting with the capture of multimodal raw data
in instrumented spaces (Section 11.3). The challenges of data annotation – mono or multimodal – are discussed in Section 11.4, while
the issues of data formatting, storage and distribution are analysed
in Section 11.5. In particular, Section 11.5.3 provides a discussion of
the access to multimodal data sets using interactive tools, and more
specifically of some meeting browsers that were recently developed
to access multimodal recordings of humans in interaction.

The chapter starts, however, by introducing in the following
Section (11.2) an important conceptual distinction between metadata
and annotations, and then surveying a number of important projects
that have created large multimodal collections, and from which best
practice examples will be drawn throughout the chapter.

11.2 SETTING THE STAGE: CONCEPTS AND
PROJECTS

Multimodal signal processing generally applies to data involving human communication modalities in two different types of

settings: human–human vs. human–computer interaction (HCI). In the first setting, the goal of multimodal processing is to grasp and abstract certain aspects of human interaction. Processing can occur either in real time, or off-line, after the interaction took place, for instance to facilitate search in multimodal recordings, or to draw conclusions regarding human performance in recorded sessions. In the second setting (HCI), the goal of multimodal processing is mainly to react in real time to a human's multimodal input to a computer, by generating an appropriate reaction (behaviour) from the computer. Although multimodal data is certainly not absent from research and development in the HCI setting (e.g., as recordings of Wizard-of-Oz experiments aimed at modelling users' behaviour), it is mainly in the first area that large quantities of annotated multimodal data are especially useful, for training and testing machine-learning tools.

In this chapter, we use the term *corpus* to refer to a meaningful set or collection of data. A multimodal corpus is a set of data files (raw data, metadata and annotations) containing recordings of humans in interaction according to several modalities. The term 'corpus' has been initially used to define a set of texts put together for a meaningful purpose, and therefore exhibiting a certain form of coherence (of topic, of style, of date, etc.). While the fields of speech and language processing commonly use corpora in their data-driven investigations, other fields simply refer to *data sets* or *collections*. Several recent initiatives (Section 11.2.2) use, however, the term 'corpus' to refer to their multimodal data sets, and corpus distributors such as the Linguistic Data Consortium have on offer a significant set of multimodal corpora. A series of workshops on this topic exists since 2000 (see http://www.multimodal-corpora.org).

11.2.1   Metadata versus Annotations

Throughout this chapter, we make reference to an important distinction: we define *annotations* as the time-dependent information which is abstracted from input signals, and which includes low-level mono or multimodal features, as well as higher-level phenomena, abstracted or not from the low-level features. For instance, speech transcription, segmentation into utterances or topical episodes, coding of gestures, visual focus of attention or dominance will be called here annotations.

Conversely, we define *metadata* as the static information about an entire unit of data capture (e.g., a session or a meeting), which is not involved in a time dependent relation to its content, i.e., which is generally constant for the entire unit. For instance, examples of metadata items are date, start and end time, location, identification of participants and indication of the media and other files associated to the unit of recording.

The terms metadata and annotations have been used with some variation depending on the field of study, but a unified terminology is important for an integrated perspective such as the one presented here. The field of speech and language studies tends to refer systematically to annotations, although speech transcript (manual or automatic) is not always explicitly considered as an annotation of the speech signal. Image and video processing tend to call metadata the features extracted from the signals, which, in the case of still images, is still coherent with the definition adopted here. In the MPEG-7 proposals, the more generic term *descriptor* conflates the two types that are distinguished here [1].

11.2.2　Examples of Large Multimodal Collections

This chapter draws from recent work involving the creation of large amounts of multimodal data accompanied by rich annotations in several modalities, and it will often quote examples from recent projects dealing with the processing of multimodal data, such as AMI/AMIDA, CHIL, M4 and IM2 projects[1].

Most examples will be drawn from the AMI Meeting Corpus [2, 3], which is one of the most recent achievements in the field of large multimodal annotated corpora, together with the smaller and less annotated CHIL Audiovisual Corpus [4]. The AMI and CHIL corpora build on experience with data from previous projects, such as

---

1. AMI/AMIDA EU integrated projects (Augmented Multiparty Interaction with Distance Access): http://www.amiproject.org, CHIL EU integrated project (Computers in the Human Interaction Loop): http://chil.server.de, M4 EU project (Multimodal Meeting Manager): http://www.dcs.shef.ac.uk/spandh/projects/m4/, IM2 Swiss National Center of Competence in Research (Interactive Multimodal Information Management): http://www.im2.ch.

the ISL Meeting Corpus [5], the speech-based ICSI Meeting Recorder corpus [6] or the M4 Corpus [7].

Recording of multimodal data had, of course, started much earlier than the aforementioned projects, but the resulting corpora (if constituted as such) were often smaller and lacked the annotations that constitute the genuine value of data resulting from the projects quoted above. Other recent multimedia initiatives focus less on annotated data, and therefore have less challenges to solve for annotation: for instance, the TRECVID collection is used to evaluate the capacity to identify a limited number of *concepts* in broadcast news (audio-video signal), from a limited list, but the reference data includes no other annotation or metadata [8]. Many more contributions can be added if one counts also the work on multimodal interfaces, as for instance in the Smartkom project [9]. However, data management is a less prominent issue in the field of multimodal HCI, as data-driven research methods seem to be less used, at least until now.

11.3   CAPTURING AND RECORDING MULTIMODAL DATA

11.3.1   Capture Devices

The capture of multimodal corpora requires complex settings such as instrumented lecture and meeting rooms, containing capture devices for each of the modalities that are intended to be recorded, but also, most challengingly, requiring hardware and software for digitising and synchronising the acquired signals. The resolution of the capture devices – mainly cameras and microphones – has a determining influence on the quality of the resulting corpus, along with apparently more trivial factors such as the position of these devices in the environment (lighting conditions, reverberation or position of speakers).

The number of devices is also important: a larger number provides more information to help defining the ground truth for a given annotation dimension. Subsequently, this annotation can serve to evaluate signal processing over data from a subset of devices only, i.e. to assess processing performance over 'degraded' signals. For instance, speech capture from close-talking microphones provides a signal that can be transcribed with better accuracy than a signal from a table-top microphone, but automatic speech recognition over the latter signal is a more realistic challenge, as in many situations people would not use headset microphones in meetings.

In addition to cameras and microphones, potentially any other sensor can be used to capture data for a multimodal corpus, though lack of standardisation means that fewer researchers will be able to work with those signals. For instance, the Anoto® technology captures handwritten notes (as timed graphical objects), while eBeam® is a similar solution for whiteboards. Presentations made during recording sessions can be recorded for instance using screen-capture devices connected to video projectors, as in the Klewel lecture acquisition system (see http://www.klewel.ch). A large number of biological sensors can capture various states of the users, from fingerprints to heart rate, eye movement or EEG. Their uses remain, however, highly experimental, because the captured data is often not general enough to be largely shared.

11.3.2   Synchronisation

Synchronisation of the signals is a crucial feature of a truly multimodal corpus, as this information conditions the possibility of all future multimodal studies using the corpus. Of course, the temporal precision of this synchronisation can vary quite a lot, the best possible value being the sampling rate of the digital signals.

Although a primitive form of synchronisation can be achieved simply by timing the beginning of recordings in each modality, there is no guarantee that the signal will remain time-aligned during the session, e.g., for 1 h or more. Therefore, a common timing device is generally used to insert periodically the same synchronisation signal in all captured modalities. For illustration purposes, this can be compared to filming the same clock on several video signals, but in reality the digital output of the synchronisation device – such as a Motu Timepiece® producing a MIDI Time Code – is embedded in each of the signals, and most accurately in each sample of a digitised signal.

The synchronisation accuracy is thus a defining feature of a multimodal corpus, and signals that are included in a corpus but with a lower synchronisation accuracy face the risk to be ignored in subsequent uses of the data. One could argue that multimodal corpora lacking suitable synchronisation information should better be called *plurimodal* rather than truly multimodal. Many of the more exotic capture devices raise difficulties in inserting synchronisation signals in their data, as is the case with the Anoto pen technology, which first stores recorded signals in a proprietary format on the pen itself.

11.3.3   Activity Types in Multimodal Corpora

Recording multimodal data collections requires a precise specification of the actions performed by the human subjects. Throughout this chapter, we refer mainly to corpora of recorded human interactions, which can range from highly constrained settings in which participants behave more or less following a well-defined scenario, to highly natural ones, in which participants interact as if no recordings were taking place (ideally even unaware of the capture devices). Each of these approaches has its challenges in terms of finding subjects and setting the stage, as well as privacy issues; however, the setting fundamentally influences the future usability of the data. Therefore, a compromise between these constraints is generally found by the corpus creators.

Multimodal corpora are also collected in settings that do not involve human interaction, such as data for multimodal biometric authentication [10], in which case naturalness of (high level) behaviour is often irrelevant. Such databases differ significantly from the ones considered here, especially in their temporal dimension.

11.3.4   Examples of Set-ups and Raw Data

Gathering a large number of capture devices in a single place, with the purpose of capturing and recording multimodal data, has given rise to *instrumented meeting or lecture rooms*, also called *smart rooms*. Several large corpora were recorded in such spaces, in many cases using several physical locations set up with similar technical specifications. For instance, the AMI smart meeting rooms [11] were duplicated in three locations, as were the CHIL lecture and meeting rooms [12]. The NIST meeting room [13] was designed to acquire data mostly for the Rich Transcription evaluations. An example of recording procedure, for the Idiap meeting room used for the AMI and IM2 projects, is described in [14, pp. 1–6].

The AMI Meeting Corpus includes captures from close-talking microphones (headset and lapel), from a far-field microphone array, close-up and room-view video cameras and output from a slide projector and an electronic whiteboard. During the meetings, the participants also used weakly synchronised Logitech Anoto pens that recorded what they wrote on paper. All additional documents, such as e-mails and presentations, produced by the participants in series of meetings were collected and added to the corpus. The meetings were recorded in English using the three AMI rooms and include mostly non-native English speakers.

The AMI Corpus[2] consists of 100 h of recordings, from 171 meetings. Most of the meetings (138 meetings, ca. 72 h) are scenario-based, made of series of four meetings in which a group undertakes a design task for a remote control prototype [2, 3]. The remainder of the meetings (33 meetings, ca. 28 h) are non-scenarised ones, mainly involving scientific discussions in groups of 3–5 people. As a comparison, the CHIL Corpus comprises 46 lectures and 40 meetings, for a total length of about 60 h, while the ICSI Meeting Corpus (audio only) consists of 75 one-hour recordings of naturally occurring staff meetings.

Even if recording such large quantities of data has specific difficulties – like equipping smart rooms and eliciting data from groups of human subjects – the most costly challenge becomes visible only after the raw data was recorded: annotating a sizable part of the data. The usefulness of a corpus is not only determined by the amount of recorded data, but especially by the amount of annotated data.

11.4   REFERENCE METADATA AND ANNOTATIONS

In this section, we discuss the process of encoding metadata and annotations once the raw data files have been captured and stored. Given our distinction between metadata and annotations, it is clear

---

2. Publicly available from http://corpus.amiproject.org.

that metadata should take much less time to encode than the reference annotations, which, being time-dependent, are potentially also very time-consuming for human annotators (e.g., speech transcription takes between 10 and 30 times real time). Although many tools and formats already exist for annotating multimodal data, the normalisation of metadata is a much less explored topic, and as a result metadata for multimodal corpora is seldom encoded explicitly.

11.4.1   Gathering Metadata: Methods

Collection of metadata is often done at the time of recording, and despite the importance of this relatively small amount of information, the collection is often non-systematic and follows ad-hoc procedures that lead to incomplete records, which are nearly impossible to recover at a later stage. This is probably due to the fact that this information is heterogeneous in nature, and seems quite minor for each meeting considered individually. It is only when the collection of multimodal recordings is gathered that the role of the metadata in organising the collection becomes obvious. These remarks apply of course to the content of the metadata information – its specific format can always be changed a posteriori.

It is important for the future usability of a collection, and for integrating it in larger pools of resources (Section 11.5.2), that metadata is gathered and consolidated into declarative files, with a content that is as detailed as possible, and a format that is as easily interpretable as possible. It is recommendable to follow existing and principled guidelines (also called norms or standards) for metadata encoding, to avoid developing a metadata specification from scratch. Well-designed guidelines minimise the number of mandatory fields, and allow for user-defined fields, in which information not foreseen in the guidelines can be encoded. It is not our purpose here to recommend particular guidelines, but some important examples are quoted below.

- MPEG-7 provides considerable detail for declaring low-level properties of the media files [1];
- Dublin Core was extended by the Open Language Archives Community (OLAC) to provide description entries mainly for speech-based and text-based data, with a relatively small

number of descriptors, which can nevertheless be extended using specifiers [15];

- NXT, the NITE XML Toolkit [16, 17], handles implicitly part of the metadata with the annotation files (see next section);
- the IMDI guidelines [18, 19], designed by the Isle MetaData Initiative, are intended to describe multimedia recordings of dialogues[3]. The format offers a rich metadata structure: it provides a flexible and extensive schema to store the defined metadata either in specific IMDI elements or as additional key/value pairs. The metadata encoded in IMDI files can be explored with the BC Browser [20], a tool that is freely available and has useful features such as search and metadata input or editing.

11.4.2 Metadata for the AMI Corpus

For the AMI Corpus, the metadata that was collected includes the date, time and place of recording and the names of participants (later anonymised using codes), accompanied by detailed sociolinguistic information for each participant: age, gender, knowledge of English and other languages, etc. Participant-related information was entered on paper forms and was encoded later into an ad-hoc XML file. However, the other bits of information are spread in many places: they can be found as attributes of a meeting in the NXT annotation files (e.g., start time), or in NXT 'metadata' files, or they are encoded in the media file names (e.g., audio channel, camera), following quite complex but well-documented naming conventions.

Another important type of metadata are the informations that connect a session, including participants, to media files such as audio and video channels, and the ones that relate meetings and documents. The former type is encoded in XML files and in file names, while the latter is only loosely encoded in the folder structure. In addition, pointers to annotation files related to each session, accompanied by brief descriptions of the annotations, should also be included in the metadata[4].

---

3. See also http://www.mpi.nl/IMDI/tools.
4. Note that there are very few attempts to normalise the descriptors of the annotation files. Recent efforts to propose a unified description language for linguistic annotations were made by the ISO TC37/SC4 group [21].

To improve accessibility to the AMI Corpus, the AMI metadata was later gathered into declarative, complete files in the IMDI format [22]. Pointers to media files in each session were first gathered from different XML resource files provided with the corpus. An additional problem with reconstructing such relations (e.g., finding the files related to a specific participant) was that information about the media resources had to be obtained directly from the AMI Corpus distribution website, because the names of media resources are not listed explicitly in the annotation files. This implies using different strategies to extract the metadata: for example, stylesheets are the best option to deal with the aforementioned XML files, while an HTTP-enabled 'crawler' is used to scan the distribution site. In addition, a number of files had to be created manually to organise the metadata files using IMDI corpus nodes, which form the skeleton of the corpus metadata structure and allow its browsing with the BC Browser.

The application of the metadata extraction tools described generated the explicit metadata for the AMI Corpus, consisting of 171 automatically generated IMDI files (one per meeting). The metadata is now available from the AMI distribution website, along with a demo access to the BC Browser over this data.

11.4.3   Reference Annotations: Procedure and Tools

The availability of manual annotations greatly increases the value and interest of a multimodal corpus. Manual annotations provide a reference categorisation of the behaviour of human subjects (in the most general sense) according to a given taxonomy or classification, in one modality or across several ones. These annotations are typically used for (a) behaviour analysis with descriptive statistics, helping to better understand communicative behaviour in a given dimension; and (b) for training and testing signal processing software for that respective dimension[5].

Reference annotations are done by human judges, using annotation tools that may include automatic processing. Annotations can be

---

5. Strictly speaking, the terms *training* and *testing* are mainly used for systems that are capable of statistical learning; but in fact any type of software will require data for development or optimisation, as well as reference data for evaluation.

completed long after the data itself was captured, and they can be made by several teams. In layered approaches such as NXT, certain annotations are based upon other ones, a fact that sets constraints on the execution order and requires proper tools. Providing automatic annotations of a phenomenon has also potential utility, either as a sample of the state-of-the-art, or when manual annotation is not feasible for a large amount of data. The nature and resolution of the initial capture devices set conditions on what phenomena can be annotated, and with what precision.

Annotations have been done since data-driven methods and quantitative testing were first used, and therefore best practice principles are at least implicitly known for each modality and phenomenon. Such principles are related to the specific definition of the phenomenon to be annotated, the training of human annotators, the design of tools, the measure of annotators' reliability (often as inter-coder agreement) and the final validation of the result (e.g., by adjudicating the output of several annotators). However, most of these annotation stages raise various scientific issues, and the needs for an *annotation science* have recently been restated in [23, pp. 8–10] for speech and language corpora.

A very large number of dimensions have been annotated in the past on mono and multimodal corpora. To quote only a few, some frequent speech- or language-based annotations are speech transcript, segmentation into words, utterances, turns or topical episodes, labelling of dialogue acts and summaries; among video-based ones are gesture, posture, facial expression or visual focus of attention; and among multimodal ones are emotion, dominance and interest-level. Many annotation tools are currently available: some are specific to a given modality, while others enable the annotation of multimodal signals. Many tools are configurable to enable annotation using a categorisation or tag set provided by the organisers of the annotations. A valuable overview of speech, video and multimodal annotation tools appears in [24], with a focus on nine currently used tools. For annotating multimodal signals, among the most popular tools are NXT, the Observer®, ANVIL [25] or Exmaralda [26].

For the AMI Corpus, annotators followed dimension-specific guidelines and used mainly but not exclusively NXT to carry out their task, generating annotations in NXT format (or similar ones) for 16 dimensions [27, 16]. Taking advantage of the layered structure of NITE annotations, several of them are constructed on top of lower-level ones. Using the NXT approach makes layered annotations consistent along the corpus, but renders them more difficult to use without the NITE toolkit. Because of the duration of the annotation process, not all the AMI Corpus was annotated for all dimensions, but a core set of meetings is available with complete annotations. The 16 dimensions are as follows: speech transcript, named entities, speech segments, topical episodes, dialogue acts, adjacency pairs; several types of summary information (abstractive and extractive summaries, participant specific ones and links between them); and in modalities other than language, focus of attention, hand gesture and head gesture. Three argumentation-related items were also partially annotated. More detailed quantitative information about each dimension is available with the corpus distribution at http://corpus.amiproject.org.

11.5   DATA STORAGE AND ACCESS

This section discusses issues of data, metadata and annotation storage and distribution, starting with hints about exchange formats, then continuing with requirements about data servers and concluding with a real-time client/server solution for accessing annotations and metadata.

11.5.1   Exchange Formats for Metadata and Annotations

Reusability is an important requirement for a multimodal corpus, given the large costs involved when creating such a resource. Therefore, the file formats that are used should be as transparent as possible, a requirement that applies to media files as well as to annotations and metadata. For media files, several raw formats or lossless compression solutions are available, and choosing one of them is often constrained by the acquisition devices, or by the resolution and sampling requirements.

Metadata and annotation formats are also generally associated to the tools that helped to encode them. However, when using annotations, it is often the case that the original format must be parsed for input to one's own signal processing tools. Given the variety of annotation dimensions, few exchange formats that are independent of annotation tools have yet been proposed. For instance, the variety of annotation formats for the language modality is visible in the MATE report [28], and more recent efforts within the ISO TC37/SC4 group have aimed at the standardisation of language-based and multimodal dialogue annotation [21], through a repertoire of normalised data categories[6].

A solution for converting the annotations of the AMI Corpus into other usable formats was put forward in [22]. The initial goal of this solution was to convert the NXT XML files into a tabular representation that could be used to populate a relational database, to which browsing tools could connect (Section 11.5.3). This conversion process can be easily modified to convert XML annotations to simpler file representations that can be used more easily by other software.

The conversion of the AMI NXT annotations proceeds as follows. For each type of annotation, associated to a channel or to a meeting, an XSLT stylesheet converts each NXT XML file into a tab-separated file, possibly using information from one or more other annotations and from the metadata files. The main goal is to resolve the NXT pointers, by including redundant information into the tables, so that each tabular annotation file is autonomous, which later speeds up queries to a database by avoiding frequent joins. Upon batch conversion, an SQL script is also created, to load the data from the tab-separated files into a relational database of annotations. This conversion process is automated and can be repeated at will, in particular if the NXT source files are updated or the tabular representation must be changed.

For metadata, a similar process was defined, to convert the IMDI files – gathered from scattered or implicit information as explained in Section 11.4.2 – to a tabular format ready for a relational database. XSLT stylesheets were defined for this conversion, and the script

---

6. Note that W3C's EMMA markup language – Extensible MultiModal Annotation – recently promoted as W3C Recommendation, is not a corpus annotation standard, but helps to convey content, in a multimodal dialogue system, between various processors of user input and the interaction manager.

applying them also generates an SQL loading script. Again, the stylesheets can be adapted as needed to generate various table formats.

11.5.2   Data Servers

Multimodal corpora being collections of data/metadata/annotation files, can be distributed as any set of files, using any digital support that appears to be convenient: CDROM, DVD or even hard disks shipped from users to provider and back to users. Of course, these files can also be distributed via a network, from a file server, which can be the main storage server or not. Another approach to accessing these resources is via automated procedures that access one item at the time, on demand, as in a search application. These approaches are briefly discussed in this section.

An example of storage and distribution server for multimodal corpora is the MultiModal Media file server, MMM, based at the Idiap Research Institute, and hosting data for the AMI, IM2, M4 and other projects. The physical architecture of the server is described in [29], while the upload/download procedures are explained in [14]. The server offers secure storage for about 1 TB of data (the majority of the space is occupied by the video and audio files) and an interface that simplifies download by building for each request a multifile download script that is executed on the user's side. For larger sets of media files, users must send hard disks that are shipped back to them with the data. The MMM server also manages personalised access rights for the different corpora.

In some cases, software components that use multimodal annotations and metadata do not need batch annotation files of past data, but rather need to query those annotations to retrieve specific items satisfying a set of conditions. These items can concern past data, but this case can be generalised to include situations when annotations produced by some signal processing modules are immediately reused by other modules. Four examples of tools that allow specific querying over data/metadata/annotations are particularly relevant here.

1. A solution for accessing metadata from search interfaces was put forward by the Open Archives Initiative (OAI) (see http://www.openarchives.org). In this view, corpus creators are

encouraged to share their metadata, which is harvested via the OAI-PMH protocol by service providers which act as consolidated metadata repositories, and facilitate the discovery of resources regardless of their exact location.

2. The NXT system provides tailored access to annotations via the NITE Query Language [17].

3. A solution for accessing specific annotation items upon request has been developed, in relation to the AMI Corpus, as the Hub client/server architecture [30]. The Hub is a subscription-based mechanism for real-time exchange of annotations, which allows the connection of heterogeneous software modules. Data circulating through the Hub is formatted as timed triples (time, object, attribute, value), and is also stored in a database. *Producers* of annotations send triples to the Hub, which are received by the *consumers* that subscribed to the respective types. Consumers can also query the Hub for past annotations and metadata about meetings. At present, the entire AMI Corpus was converted to timed triples and inserted into the Hub's database as past data.

4. A solution for multichannel and multitarget streaming was also developed to complement the Hub's mechanism in the domain of video and audio media files. The HMI Media Server (University of Twente, *unpublished*) can broadcast audio and video that is captured in an instrumented meeting room to various consumers, possibly remotely, thus allowing a flexible design of interfaces that render media streams.

11.5.3 Accessing Annotated Multimodal Data

Accessing multimodal data from an end-user's point of view involves the use of multimodal browsers and other tools that are capable of rendering media files and enhancing them with information derived from metadata and annotations. This section briefly introduces the concept of a meeting browser and exemplifies an application for speech-based retrieval of multimodal processed data.

The concept of meeting browsing has emerged in the past decade [31, 32], partly in relation to the improvements brought to multimodal signal processing technology. Meeting browsers generally take advantage of multimodal processing to improve users' access to meetings. This can mean either speeding up the search for a specific piece of information, or facilitating the understanding of a whole meeting through some form of summarisation.

The JFerret framework offers a customisable set of plugins or building blocks that can be hierarchically combined to design a meeting browser. Using JFerret and other standard tools, several meeting browsers have been implemented. The creators of JFerret proposed sample instantiations of the framework, mainly based on speech, video, speaker segmentation and slides [33, 34]. The JFerret framework was also used for other browsers, such as audio-based browsers, or dialogue-based, document-centric or multimodal ones in the IM2 project [35].

The Automatic Content Linking Device (ACLD) [36] demonstrates the concept of real-time access to a multimodal meeting corpus, by providing 'just-in-time' or 'query-free' access to potentially relevant documents or fragments of recorded meetings, based on speech from an ongoing discussion. The results are shown individually to meeting participants, which can examine the documents or the past meeting fragments in further detail, e.g., using a meeting browser, if they find that their contents are relevant to the current discussion.

Evaluation of access tools to multimodal data is a challenging topic, and evaluation resources such as BET [37, 38], which consist of true/false statements about a meeting produced by independent observers, could be added to the multimodal corpora as an extended form of annotation.

11.6   CONCLUSIONS AND PERSPECTIVES

The volume of multimodal data that is recorded and made available as a collection is bound to increase considerably in the near future, due to the rapid diffusion of recording technology for personal, corporate or public use, and due to the improvement of processing tools that are available – although the robust extraction of semantic primitives from multimodal signals remains a major issue.

Most of the important problems that remain to be solved concerning data, metadata and annotations are related to the challenge of increasing the interoperability of multimodal resources. The solution will involve the specification of shareable annotations, thanks to common data categories that can be further specified according to each project. Another challenge is the exchange of metadata [39] to facilitate the discovery of such resources via integrated catalogues.

This chapter has summarised the main stages of multimodal data management: design, capture, annotation, storage and access. Each stage must be properly handled so that the result constitutes a reusable resource with a potential impact on research in multimodal signal processing. Indeed, as shown in the other chapters of this book, the availability of annotated multimodal collections tends to determine the nature and extent of data-driven processing solutions that are studied.

Acknowledgments

The author would like to acknowledge the support of the AMIDA and IM2 projects (see note 1).

REFERENCES

1. B.S. Manjunath, P. Salembier, T. Sikora, Introduction to MPEG-7: Multimedia Content Description Interface. John Wiley and Sons, Hoboken, NJ, 2002.
2. J. Carletta, S. Ashby, S. Bourban, M. Flynn, M. Guillemot, T. Hain, et al., The AMI Meeting Corpus: A pre-announcement, in: S. Renals and S. Bengio (Eds.), Machine Learning for Multimodal Interaction II, LNCS 3869, Springer-Verlag, Berlin/Heidelberg, 2006, pp. 28–39.
3. J. Carletta, Unleashing the killer corpus: experiences in creating the multi-everything AMI Meeting Corpus, Lang. Resour. Eval. J. 41 (2) (2007) 181–190.
4. D. Mostefa, N. Moreau, K. Choukri, G. Potamianos, S. Chu, A. Tyagi, et al., The CHIL audiovisual corpus for lecture and meeting analysis inside smart rooms, Lang. Resour. Eval. 41 (3) (2007) 389–407.
5. S. Burger, V. MacLaren, H. Yu, The ISL meeting corpus: The impact of meeting type on speech style, in: ICSLP 2002 (International Conference on Spoken Language Processing), Denver, CO, 2002.
6. A. Janin, D. Baron, J.A. Edwards, D. Ellis, D. Gelbart, N. Morgan, et al., The ICSI meeting corpus, in: ICASSP 2003 (International Conference on Acoustics, Speech, and Signal Processing), vol. 1 (2003) pp. 364–267, Hong Kong.
7. I. McCowan, S. Bengio, D. Gatica-Perez, G. Lathoud, F. Monay, D. Moore, et al., Modeling human interaction in meetings, in: ICASSP 2003 (International

Conference on Acoustics, Speech, and Signal Processing), vol. 4 (2003) pp. 748–751, Hong Kong.

8. A.F. Smeaton, P. Over, W. Kraaij, High-level feature detection from video in TRECVid: a 5-year retrospective of achievements, in: A. Divakaran (Ed.), Multimedia Content Analysis, Theory and Applications, Springer-Verlag, Berlin, 2009, pp. 151–174.

9. W. Wahlster (Ed.), SmartKom: Foundations of Multimodal Dialogue Systems. Springer-Verlag, Berlin, 2006.

10. A. Ross, A.K. Jain, Multimodal biometrics: an overview, in: EUSIPCO 2004 (12th European Signal Processing Conference), Vienna, 2004, pp. 1221–1224.

11. D.J. Moore, The Idiap Smart Meeting Room. Communication 02-07, Idiap Research Institute, 2002.

12. H.K. Ekenel, M. Fischer, R. Stiefelhagen, Face Recognition in Smart Rooms, in: 4th Joint Workshop on Multimodal Interaction and Related Machine Learning Algorithms (MLMI 2007), Brno, 2007, pp. 120–131.

13. V. Stanford, J. Garofolo, O. Galibert, M. Michel, C. Laprun, The NIST smart space and meeting room projects: signals, acquisition annotation, and metrics, in: ICASSP 2003 (IEEE International Conference on Acoustics, Speech, and Signal Processing), vol. 4 (2003) pp. 736–739, Hong Kong.

14. M. Guillemot, B. Crettol, From meeting recordings to web distribution: Description of the process. Idiap-Com Idiap-Com-05-2005, Idiap Research Institute, 2005.

15. S. Bird, G. Simons, Extending Dublin Core metadata to support the description and discovery of language resources, Comput. Hum. 37 (4) (2001) 375–388.

16. J. Carletta, J. Kilgour. The NITE XML toolkit meets the ICSI meeting corpus: Import, annotation, and browsing, in: S. Bengio, H. Bourlard (Eds.), Machine Learning for Multimodal Interaction, LNCS 3361, Springer-Verlag, Berlin/Heidelberg, 2005, pp. 111–121.

17. J. Carletta, S. Evert, U. Heid, J. Kilgour, The NITE XML toolkit: data model and query language, Lang. Resour. Eval. 39 (4) (2005) 313–334.

18. P. Wittenburg, W. Peters, D. Broeder, Metadata proposals for corpora and lexica, in: LREC 2002 (Third International Conference on Language Resources and Evaluation), Las Palmas, 2002, pp. 1321–1326.

19. D. Broeder, T. Declerck, L. Romary, M. Uneson, S. Strömqvist, P. Wittenburg, A large metadata domain of language resources, in: LREC 2004 (4th International Conference on Language Resources and Evaluation), Lisbon, 2004, pp. 369–372.

20. D. Broeder, P. Wittenburg, O. Crasborn, Using profiles for IMDI metadata creation, in: LREC 2004 (4th International Conference on Language Resources and Evaluation), Lisbon, 2004, pp. 1317–1320.

21. H. Bunt, L. Romary, Standardization in multimodal content representation: Some methodological issues, in: LREC 2004 (4th International Conference on Language Resources and Evaluation), vol. 6 (2004) pp. 2219–2222, Lisbon.

22. A. Popescu-Belis, P. Estrella, Generating usable formats for metadata and annotations in a large meeting corpus, in: ACL 2007 (45th International Conference of the Association for Computational Linguistics), Prague, 2007, pp. 93–96.

23. L. Liddy, E. Hovy, J. Lin, J. Prager, D. Radev, L. Vanderwende, et al., MINDS workshops: natural language processing. Final report, National Institute of Standards and Technology (NIST), 2007.

24. L. Dybkjær, N.O. Bernsen, Towards general-purpose annotation tools – how far are we today? in: LREC 2004 (4th International Conference on Language Resources and Evaluation), vol. 1 (2004) pp. 197–200, Lisbon.

25. M. Kipp, ANVIL: a generic annotation tool for multimodal dialogue, in: Eurospeech 2001 (7th European Conference on Speech Communication and Technology), Aalborg, 2001, pp. 1367–1370.

26. T. Schmidt, The transcription system EXMARaLDA: an application of the annotation graph formalism as the basis of a database of multilingual spoken discourse, in: IRCS Workshop on Linguistic Databases, Philadelphia, PA, USA, 2001, pp. 219–227.

27. J. Carletta, S. Evert, U. Heid, J. Kilgour, J. Robertson, H. Voormann, The NITE XML toolkit: flexible annotation for multi-modal language data, Behav. Res. Methods Instrum. Comput. 35 (3) (2003) 353–363.

28. M. Klein, N.O. Bernsen, S. Davies, L. Dybkjær, J. Garrido, H. Kasch, et al., Supported coding schemes. MATE Deliverable D1.1, MATE (Multilevel Annotation, Tools Engineering) European Project LE4-8370, 1998.

29. F. Formaz, N. Crettol, The Idiap multimedia file server. Idiap-Com Idiap-Com-05-2004, Idiap Research Institute, 2004.

30. AMIDA, Commercial component definition. Public Deliverable D6.6 (ex D7.2), AMIDA Integrated Project FP7 IST–033812 (Augmented Multi-party Interaction with Distance Access), 2007.

31. A. Nijholt, R. Rienks, J. Zwiers, D. Reidsma, Online and off-line visualization of meeting information and meeting support. Vis. Comput. 22 (12) (2006) 965–976.

32. S. Whittaker, S. Tucker, K. Swampillai, R. Laban, Design and evaluation of systems to support interaction capture and retrieval, Pers. Ubiquitous Comput. 12 (3) (2008) 197–221.

33. P. Wellner, M. Flynn, M. Guillemot, Browsing recorded meetings with Ferret, in: S. Bengio, H. Bourlard (Eds.), Machine Learning for Multimodal Interaction I, LNCS 3361, Springer-Verlag, Berlin/Heidelberg, 2004, pp. 12–21.

34. P. Wellner, M. Flynn, M. Guillemot, Browsing recordings of multi-party inter-actions in ambient intelligent environments, in: CHI 2004 Workshop on "Lost in Ambient Intelligence", Vienna, 2004.
35. D. Lalanne, A. Lisowska, E. Bruno, M. Flynn, M. Georgescul, M. Guillemot, et al., The IM2 Multimodal Meeting Browser Family. Technical report, (IM)2 NCCR, 2005.
36. A. Popescu-Belis, E. Boertjes, J. Kilgour, P. Poller, S. Castronovo, T. Wilson, et al., The AMIDA automatic content linking device: just-in-time document retrieval in meetings, in: A. Popescu-Belis and R. Stiefelhagen (Eds.), Machine Learning for Multimodal Interaction V (Proceedings of MLMI 2008, Utrecht, 8–10 September 2008), LNCS 5237, Springer-Verlag, Berlin/Heidelberg, 2008, pp. 272–283.
37. P. Wellner, M. Flynn, S. Tucker, S. Whittaker, A meeting browser evaluation test, in: CHI 2005 (Conference on Human Factors in Computing Systems), Portland, OR, 2005, pp. 2021–2024.
38. A. Popescu-Belis, P. Baudrion, M. Flynn, P. Wellner, Towards an objective test for meeting browsers: the BET4TQB pilot experiment, in: A. Popescu-Belis, H. Bourlard, S. Renals (Eds.), Machine Learning for Multimodal Interaction IV, LNCS 4892, Springer-Verlag, Berlin/Heidelberg, 2008, pp. 108–119.
39. P. Wittenburg, D. Broeder, P. Buitelaar, Towards metadata interoperability, in: NLPXML 2004 (4th Workshop on NLP and XML at ACL 2004), Barcelona, 2004, pp. 9–16.

Part III

Multimodal Human–Computer and Human-to-Human Interaction

Chapter 12

Multimodal Input

Natalie Ruiz*, Fang Chen*, and Sharon Oviatt†
*NICTA and The University of New South Wales, Australia;
†Incaa Designs, Bainbridge Island WA, USA

12.1 INTRODUCTION

Multimodal interfaces involve user input through two or more combined modes, such as speech, pen, touch, manual gestures, gaze and head and body movements. They represent a new direction for

Multimodal Signal Processing, ISBN: 9780123748256

computer interfaces and a paradigm shift away from conventional graphical interfaces because they involve recognition-based technologies designed to handle continuous and simultaneous input from parallel incoming streams, processing of input uncertainty using probabilistic methods, distributed processing and time-sensitive architectures. This new class of interfaces aims to recognise naturally occurring forms of human language, which will be discussed in this chapter. The advent of multimodal interfaces based on recognition of human speech, gaze, gesture and other natural behaviour represents only the beginning of a progression toward computational interfaces capable of relatively human-like sensory perception.

This chapter will begin with a brief review of the advantages of multimodal interfaces and present some examples of state-of-the-art multimodal systems. It is focused on the links between multimodality and cognition, namely the application of human cognitive processing models to improve our understanding of multimodal behaviour in different contexts, particularly in situations of high mental demand. To capitalise on the power of the multimodal paradigm to facilitate effortful tasks, we must design systems that can aid task-completion when the user needs it most – during tasks that induce high cognitive load. In the last section of this chapter, we discuss how patterns of multimodal input can inform and inspire the design of adaptive multimodal interfaces and the wide ranging benefits of such adaptation. As a case-study, the educational domain presents a perfect requirement for interfaces that minimise students' cognitive load such that they are capable of supporting both high- and low-performing students. Our chapter concludes by identifying fertile research directions for multimodal research.

12.2 ADVANTAGES OF MULTIMODAL INPUT INTERFACES

The growing interest in *multimodal interface* design is inspired largely by the goal of supporting more flexible, powerfully expressive, and low cognitive load means of human–computer interaction (HCI). Multimodal interfaces are expected to be easier to learn and use and are preferred by users for many applications. They have the potential to expand computing to more mobile and complex real-world applications, to be used by a broader spectrum of everyday people and to accommodate more adverse usage conditions than in the past. Some of the more notable advantages are

- *Robustness*: Redundancy in multimodal input increases the quality of communication between the user and system because conveying similar or related information through different modalities increases the likelihood of recognition [1]. Modalities work together to achieve a greater level of expressiveness by 'refining imprecision' or modifying the meaning conveyed through another modality [1, 2]. Strategies of *mutual disambiguation* and unification-based input fusion algorithms, where integrated multimodal inputs are processed together, can help increase the system's performance.
- *Naturalness*: Multimodal communication can result in a high degree of naturalness, capitalising on the well-established practices of human–human interaction [1]. Complex tasks can be facilitated through use of multimodal interaction because the paradigm effectively increases the communicative bandwidth between the user and system, increasing the level of input expressivity.
- *Flexibility*: A major benefit of multimodal interfaces is flexibility, allowing for individual users to both perceive and structure their communication in diverse ways for specific contexts. Users can choose which modalities they employ, and how they are structured along semantic, temporal and syntactic dimensions. Users have also indicated a strong preference for *integrated multimodal* input in spatially-oriented (map-based) tasks [3]; however, they tend to intermix unimodal and multimodal interaction as they see fit [4].
- *Minimising errors*: Multimodal interfaces have been shown to increase performance by lowering the number of errors (through error avoidance) and spontaneous disfluencies, compared with speech-only interfaces, in geographical interactive tasks [5]. Although multimodal interfaces have been shown to help users achieve shorter task completion times, particularly with map-based tasks [3], compared with traditional graphic user interface or speech-only interfaces, efficiency is not considered to be one of the major advantages of the multimodal paradigm. However, users can achieve faster error-correction when using multimodal interfaces [3, 5].

12.2.1   State-of-the-Art Multimodal Input Systems

The state-of-the-art multimodal input systems are currently only able to process two to three modal input channels, are application-specific and have limited interaction vocabularies and grammars [6]. Many issues relating to synchronisation, multimodal fusion (input interpretation), multimodal fission (output presentation) and interaction design are still being explored.

The two most mature types of multimodal systems, in which the keyboard and mouse have been completely replaced, are those that combine either speech and pen input [6] or speech and lip movements [7–9]. For speech and pen systems, spoken language is sometimes processed along with complex pen-based gestural input, involving hundreds of different symbolic interpretations beyond pointing [10]. For speech and lip movement systems, spoken language is processed along with corresponding human lip movements during the natural audio-visual experience of spoken interaction. Research has been directed toward quantitative modelling of the integration and synchronisation characteristics of the two rich input modes being processed, and innovative time-sensitive architectures have been developed to process these patterns in a robust manner. A frame-based method of information fusion and *late semantic fusion* approach is employed in most speech and pen systems. Quickset used a statistically-ranked unification process and a hybrid symbolic/statistical architecture [11], and other recent systems have adopted a similar approach for processing both multimodal input and output [12–15]. A glossary of relevant terms is provided in Figure 12.1.

- *Multimodal interfaces* process two or more combined user input modes – such as speech, pen, touch, manual gestures, gaze and head and body movements – in a coordinated manner with multimedia system output. The paradigm aims to recognise naturally occurring forms of human language and behaviour.
- *Mutual disambiguation* involves disambiguation of signal or semantic-level information in one error-prone input mode from partial information supplied by another. It leads to recovery from unimodal recognition errors within a multimodal architecture, with the net effect of suppressing errors experienced by the user.

- *Simultaneous integration* refers to the pattern of two input signals (e.g., speech and pen) produced in a temporally overlapped manner. *Sequential integration* refers to the pattern of inputs where the multimodal signals are separated, one is presented before the other with a brief pause intervening.
- *Semantic-level* or *late fusion* is a method for integrating semantic information derived from parallel input modes in a multimodal architecture, which has been used for processing speech and gesture input.
- *Schemas* are internal representations of the world, or mental models that are organised in categories and stored in long-term memory.
- *Transfer* occurs when material learned in one situation assists in another; this is achieved through a process of generalisation and application into a different context.
- *Theory of working memory* was proposed by Baddeley and Hitch [16] and contends that at least two slave systems are responsible for working memory processing (visual and auditory), and a central executive integrates information and coordinates the slave systems.
- *Multiple resource theory*, proposed by Wickens [17], contends that tasks compete for various different working memory resources at each stage of processing (perception, interpretation and response processing). Interference occurs when two tasks require the same resource, hence performance degrades.
- *Gestalt theory* comprises a set of theoretical and methodological principles arising from the idea that the human brain has self-organising tendencies that cause stimuli to be perceived as coherent forms in the first instance: 'the whole is different from the sum of its parts'.
- *Cognitive load theory* is a prescriptive framework, based on theories of cognitive architecture and learning processes. Its aim is to design stimulus materials that are better suited to our cognitive function and its limitations.
- *Mutually adaptive multimodal interfaces* are ones in which both user and system co-adapt during interaction, resulting in a more predictable and synchronous whole.

Figure 12.1 Glossary of terms.

Multimodal systems that incorporate continuous 3D manual gesture modalities as well as speech have recently come to the fore; however, these are far less mature when compared to the technologies used to recognise 2D pen input and ink modalities [18, 19]. Naturally, new challenges encompassed by gestural input including segmentation, feature extraction and interpretation of spontaneous movement trajectories, need to be addressed before such systems become more prevalent. The introduction of other modalities, including modalities driven by vision-based engines, such as gaze, facial expressions, head nodding, gesturing and large body movements, is fuelling a growing interest in novel kinds of interactions, which are still in the early stages of research [19–21]. For illustrative purposes, a comparison of four different speech and gesture systems is summarised in Figure 12.2. Quickset, an early multimodal pen/voice system developed by Cohen et al. [22], is the most mature of all four. Others include Boeing's VR Aircraft Maintenance Training System, developed by Duncan et al. [23]; the PEMMI system, a demonstrator interface developed by Chen et al. at NICTA [24], which supports using freehand manual gestures together with speech, to monitor traffic and the Portable Voice Assistant, developed by Bers et al. at BNN [25].

Multimodal systems	Quickset	VR Aircraft Maintenance Training	PEMMI Traffic Monitoring System	Portable Voice Assistant
Input modalities	Pen input – large vocabulary Speech – grammar-based, moderate vocabulary	Small vocabulary speech and 3D manual gestures, grammar-based.	Small vocabulary speech, and freehand manual gestures, keyword-based	Small vocabulary speech and Pen input – deictic selection, grammar-based
Type of signal fusion	Late semantic fusion, unification, hybrid symbolic/ statistical MTC framework	Late semantic fusion, frame-based	Late semantic fusion, frame-based	Late semantic fusion, frame-based
Type of platform and applications	Wireless handheld varied map and VR applications, digital paper	Virtual reality system, aircraft maintenance training	Laptop + large screen traffic monitoring over large areas	Wireless handheld, catalogue-ordering

FIGURE 12.2 Different multimodal systems and their capabilities.

12.3 MULTIMODALITY, COGNITION AND PERFORMANCE

The close relationship between multimodality and cognition exists on a number of levels. First, psychological and cognitive theories of multimodal processing, together with recent neurological evidence, can provide a biological and evolutionary perspective to explain the advantages of multimodal interaction. The classic example is the study of interpersonal dialogue. Second, the literature reveals that cognitive processes and user performance are closely linked. The capacity of working memory can impose severe limitations on performance and learning. *Cognitive load theory* attempts to break down the relationships between stimulus materials and performance and identifies multimodal processing as a method for improving performance in tasks that induce high mental demand.

12.3.1 Multimodal Perception and Cognition

Human beings are physiologically designed to acquire and produce information through a number of different modalities: the human communication channel is made up of sensory organs, the central nervous system, various parts of the brain and effectors (muscles or glands) [26]. Sensory inputs from specific modalities each have their own individual pathway into a primary sensory cortex and can be processed in parallel [26]. In addition, there are specific multimodal integration (input) and diffusion (output) association areas in the brain that are highly interconnected [26]. Adding credence to the raft of empirical evidence, neuroimaging technology such as positron emission topography and functional magnetic resonance imaging have been used to identify separate locations for the verbal/auditory, imagery/spatial and executive functions of working memory [27, 28]. Multimodal perception and cognition structures in the human brain appear to have been specifically designed to collate and produce multimodal information. This is nowhere more apparent than in studies of interpersonal dialogue.

The multimodal aspects of conversational speech and gesture have received much attention in the literature. Communication is inherently multimodal: we talk to one another, wave our arms in the air and make eye contact with our dialogue partners, who also perceive, interpret and understand the protocols of conversational dialogue [29, 30]. Although speech and gesture channels do not always convey the same information, it is always semantically and pragmatically compatible [30]. The most prevalent and widely accepted theory on the relationship between these modes of communication sees speech and gesture production as an integrated process, generated from a common underlying mental representation. Both modes are therefore considered to be equally functional in creating communicative meaning [29]. In this way, human communication can be seen to exploit our natural ability to easily process and produce multimodal information. Such a principle can be directly applied to the design and implementation of multimodal interfaces by allowing users to make flexible use of the entire gamut of modal productions (e.g., gaze, gesture and speech).

12.3.2   Cognitive Load and Performance

The relationship between working memory and performance is explored by Sweller's *Cognitive load theory*. The theory is driven by empirical observations of how well people are able to learn from different stimulus materials and corresponding hypotheses based on well-established modal models of working memory architectures such as Baddeley's model [16] (see Section 12.4.1 for details).

Cognitive load is a concept that attempts to describe the experience of mental demand, adding an interesting dimension to performance assessment. The theory rests on the assumption that working memory is limited in capacity [31] and duration [32]. Tasks with very high or very low levels of cognitive load can severely impact a subject's performance: if too high, the subject will not have sufficient resources to perform well, and if too low, there is a chance that the subject is not being cognitively engaged in an optimal way [33]. Hence, effective use of working memory processing is vital for achieving successful knowledge *transfer*. Subjects exhibiting similar levels of performance also may differ in their individual cognitive load experience. The theory describes three types of cognitive load that contribute to mental demand [34]:

- *Intrinsic load*: It refers to the cognitive load created by the conceptual complexity of the task and cannot be easily changed [34].
- *Extraneous load*: It is the load created by the representation of the task and can be easily manipulated by changing the format of the presentation modalities and design [34].
- *Germane load*: It is said to be caused by the conversion of novel information into *schema* and then stored in long-term memory [34].

The connection between cognitive load and multimodal processing can be traced back to an early recurring observation reported in the literature, dubbed the 'modality effect'. It was found that if two different modalities were used in task stimuli, subjects exhibited both *performance advantage* and *preference* for multiple modality conditions, compared with single modality tasks. Cognitive load theory addresses this observation by attributing the advantages of multimodal processing to a set of modality-specific working memory resources that effectively expand working memory when multimodality is used [35]. The theory assumes a working memory model that has semi-independent modules for processing visual/spatial and verbal/auditory information and a central executive which interacts with long-term memory stores [16, 17].

Furthermore, evidence shows users are known to change and adapt their multimodal behaviour in complex, high-load situations. For example, when tasks are more difficult, users prefer to interact multimodally rather than unimodally across a variety of different application domains. It is believed that this facilitates more effective use of modality-based working memory resources and assists users in self-managing cognitive load [35–40].

12.4   UNDERSTANDING MULTIMODAL INPUT BEHAVIOUR

Modal theories or frameworks of working memory processes can account for phenomena that occur on the user-side of multimodal system processing, both in terms of input and response production. Interface designers can leverage cognitive science contributions in a number of ways. From an engineering perspective, an understanding of the way multimodal information is produced can aid the design and implementation of improved signal recognition and signal fusion algorithms, developing more appropriate strategies for *real-time adaptation* features in those technologies. In terms of content and representational design, more effective output presentation strategies can be developed, to optimise the user's natural information acquisition processes. The evaluation of existing interfaces can also be judged according to how well they are able to cater and adapt to naturally occurring multimodal input patterns.

12.4.1 Theoretical Frameworks

The models reviewed below describe the relationship between the goal of accomplishing a cognitive task and the allocation of a subject's internal resources from a modal perspective. They provide a reference point to begin the analysis of multimodal behaviours.

Modal Theory of Working Memory

Baddeley's *theory of working memory* [16] suggests that working memory is composed of independent modal processors that work together in a coordinated and synchronous fashion. Baddeley and colleagues draw evidence from patient pathologies, neural imaging and empirical findings, incorporating these into a single model, which is composed of at least four main components:

- *The central executive*: Allocates subtasks to a modal processor and coordinates interaction between the modal slave systems;
- *The phonological loop*: The first of three slave systems, it exclusively processes verbal, auditory and linguistic tasks;
- *The visuo-spatial sketchpad*: Exclusively processes imagery and spatial tasks and
- *The episodic buffer*: Temporarily stores schema retrieved from long-term memory and representations created by the executive or either of the two other slave systems.

The popularity of Baddeley's model lies in its ability to account for much of the psychological evidence available as well as neurological pathologies. The model can offer a high-level view of multimodal input processes, indicating the types of resources that are likely to be used in generating specific productions. For example, speech inputs are likely generated in the auditory/verbal processors, whereas gestural input is processed by the visual/spatial component.

However, despite recent subcomponent additions to the model, such as the *visual cache* within the *sketchpad* [41], it must be acknowledged that Baddeley's initial model somewhat oversimplifies the distinction between the visual/spatial and auditory/verbal function, rendering the model unable to explain combined or interleaved cognitive processes that may involve both systems, such as those that occur in multimodal productions, or the spatial and linguistic characteristics of writing.

Multiple Resource Theory

At the other end of the spectrum, the motivation behind *multiple resource theory* is to predict the level of performance and productivity expected of a human operator carrying out multiple real-world tasks [17] and the compatibility between the modalities used for input and output processing. The model's primary advantage is that it accounts for performance effects in highly taxing *applied task* combinations, such as driving and speaking on the phone or monitoring aircraft visually and attending to incoming auditory messages.

Multiple resource theory describes a set of modally organised central resources that are taxed when a user completes a task. These resources have limited capacity and can be shared by multiple tasks. *Task interference* occurs when two concurrent tasks requiring the same resource compete or interfere with one another, causing performance on both tasks to degrade. Interference is predicted if the same modal resources are required by more than one task at any stage of the cognitive process. Conflicts do not exist within a single task, as processing is assumed to occur sequentially within a single task [17].

This theory can be useful in predicting whether modal interference will occur in a multiple-task situation and ultimately provide an estimate of the performance of the operator. However, there is less detail in the theory about the processes involved in sharing these resources, e.g., modal sharing within a single task, where all processes are semantically dependent on each other, as often occurs in the generation of multimodal input.

Gestalt Theory

Although *Gestalt theory* has been used to explain human perception of visual-spatial effects, the principles have also been applied to the perception of acoustic, haptic and other sensory information [42]. More recently still, research has explored the relevance of these features within multimodal productions (input) in multimodal interfaces [43].

All Gestalt principles rest on the creation of a balanced and stable perceptual form that is capable of maintaining its equilibrium, through a reciprocal action and reaction of internal and external physical forces [44]. The principles related to grouping information as a 'coherent whole' are the most well known in the theory and widely appropriated [44]. Studies of multimodal interaction patterns have used these principles to shed new light on communicative structures. For example, multimodal pen/voice interactions comprise individual modal inputs that convey meaning within their temporal and spatial features. Gestalt theory's principle of proximity states that spatial or temporal proximity can cause elements to be perceived as related. Interpreted in this fashion, speech- and pen-based, possibly much like speech- and gesture-based communicative structures, may be organised along a common temporal dimension thus providing implicit cues that prescribe the receiver's interpretation of the signals. It follows that changes in the temporal arrangement of these modalities can represent varying semantics related to proximity and relatedness in input and output constructions [43].

The Gestalt principles of symmetry and area [44] have also been applied to multimodal input patterns [43]. Instances of pen/voice multimodal inputs exhibiting symmetrical features include multimodal structures with close temporal correspondence between individual modal signals, e.g., tight coupling in the co-timing of multimodal signal onsets and offsets [43]. The Gestalt principle of area describes the tendency to group elements in such a way as to minimise the size of the visible outline or the shortest temporal interval. Such a principle may translate into a notable propensity for users to communicate using *simultaneous integration* rather than *sequential integration* of multimodal signals, in a strategy of temporal economy, or perhaps a higher frequency of simultaneous to sequential constructions in general [43].

12.4.2   Interpretation of Multimodal Input Patterns

An array of different multimodal input patterns, from both active and passive modes, have been analysed by researchers in attempts to gain some level of insight into the cognitive processes of the user. As has been discussed so far, it has been shown that multimodal processing (and input patterns) can be affected by high cognitive load. Users experiencing high load will behave – and therefore, interact – differently compared with users experiencing low load. This is due to the intricate relationships that exist between use of multimodality, cognition and performance. The evidence shows that users change and adapt their multimodal behaviour to increase their performance, both in terms of temporal efficiency and accuracy [36, 40]. For example, Kendon suggests that ideas that are difficult to express in language are communicated through accompanying gesture [45]. Elsewhere it has been suggested that some types of gestures decrease cognitive load during an explanation task [37]. Users also tend to have a higher number of disfluencies overall during human–computer interaction (HCI) when completing tasks with high load (e.g., utterances that contain locative content vs. utterances that do not [36]).

It has also been shown that people tend to spread information acquisition [35] and production [36] processes over distinct modalities during complex tasks. As task complexity and cognitive load increases, users tend to react by being *more multimodal*. Multimodal communicative strategies appear to change in such a way as to maximise the usage of Baddeley's modal working memory resources (as discussed in Section 12.4.1) and reduce the potential for Wickens' notions of interference (as discussed in Section 12.4.1). For example, one study revealed multimodal interaction is most likely to be used when the salience of the object being selected is low [40]. This is arguably a more complex task than selecting a highly salient object:

using more than one modality would effectively disseminate the load to relevant processing units in working memory. The study also suggests that speech and pen input sets will consistently carry complementary information in each modal signal [46], a strategy which results in similar efficacy outcomes. This change to multimodal input production may be due to a strategy of self-management of cognitive load, a hypothesis that is consistent with the modal frameworks presented above.

Empirical evidence also shows that semantic features of multimodal input are sensitive to increases in task complexity and cognitive load [47]. People deal with complex tasks by channelling the required semantic and representational processing to exclusive modal resources, with the least amount of replication possible, resulting in semantically divergent modal inputs [48]. This strategy causes decreasing redundant and increasing complementary patterns in multimodal communicative structures [47]. The example shown in Figure 12.3 depicts the symbolic differences between producing a redundant message and a complementary message, using gesture and speech inputs. It is presented in the framework of Baddeley's model of working memory and illustrates the areas taxed by processing load. Complementary input patterns represent a strategy of cognitive economy during high-load tasks over redundant input patterns [47].

Changes due to high load characteristic to each modality are also in evidence. For example, in speech input specifically, significant variations in levels of spoken disfluency, articulation rate, and filler and pause rates have been observed in subjects during low versus high cognitive loads, reflecting the difficulty experienced in executing responses to harder tasks [49]. Pause-based features such as pause length, rate of pauses and total time pausing have been found to increase from low load to high load in spoken comprehension tasks and speech interaction tasks [50, 51]. This behaviour suggests some level of strategic execution of communicative signals, where the subject allows themselves more time to think while speaking during the execution of utterances with complex content. The data collected in other studies of multimodal behaviour suggest that surface-level multimodal features in speech and pen-gesture input will betray symptoms of high load in some user groups, with the quality of the signal degrading significantly [47]. This may be due to the unavailability of

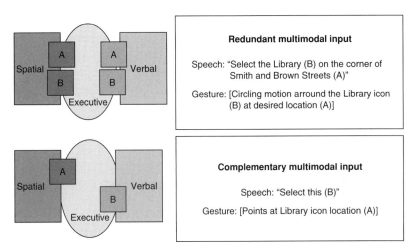

FIGURE 12.3   Processing of redundant and complementary multimodal input.

resources and subsequent scant allocation to attend to the planning and execution of individual modal signals, specifically in tasks where the conceptual complexity is high [47].

Multimodal interfaces are, in themselves, known to reduce the level of cognitive load over tasks completed using unimodal interfaces. Evidence of these input patterns during multimodal interaction may provide implicit indications that a user is experiencing high load. Systems equipped with methods for unobtrusive, real-time detection of cognitive load and general cognitive load awareness will be able to adapt content delivery in a more intelligent way by sensing what the user is able to cope with at any given moment. Presentation and interaction strategies can be used to adapt the pace, volume and format of the information conveyed to the user.

12.5   ADAPTIVE MULTIMODAL INTERFACES

Recent advances in the design of applications and user interfaces have promoted the awareness of the user context as well as user preferences. As discussed in the previous section, the cognitive load of a user represents an important factor to be considered in adaptive human–computer interfaces, especially in scenarios of high intensity work conditions and complex tasks.

Adaptive processing is one of the major future directions for multimodal interface design. Multimodal interfaces need to be developed that can adapt to individual users and also adapt continuously to user interaction with a system. Adaptive processing is especially important for multimodal interface design for several reasons, including the suitability of multimodal interfaces for

- Dynamically changing mobile environments, which could benefit from adapting to context;
- Complex tasks in which users' cognitive load fluctuates from low to high, which could benefit from adapting to users' current load level;
- Learning-based tasks involving support for many representational systems, which could benefit from adapting to different representations as students express themselves during problem solving steps and
- Universal access to computation and information resources by diverse populations (e.g., handicapped users, indigenous and minority users), which could benefit from adaptation to an individual user, their characteristics (e.g., expertise level, language), current status (e.g., focus of attention) and needs within a given domain (e.g., sensory or cognitive deficits).

12.5.1 Designing Multimodal Interfaces that Manage Users' Cognitive Load

The development of computer systems historically has been a technology-driven phenomenon, with technologists believing that 'users can adapt' to whatever they build. Human-centred design advocates that a more promising and enduring approach is to model users' natural behaviour at the start, including any constraints on their ability to attend, learn and perform, so that interfaces can be designed that are more intuitive, easier to learn and freer of performance errors. By minimising cognitive load, users' performance and safety especially can be enhanced in field and mobile situations and when working on more complex tasks. Examples of basic multimodal design strategies that are capable of minimising users' cognitive load and thereby improving performance include

1. *Modelling users' natural multimodal communication patterns*: People are experienced multimodal communicators during interpersonal interaction (Section 12.3.1). Their language also typically is simpler and easier to process when communicating multimodally. By modelling users' existing language, rather than attempting to retrain strongly entrenched patterns (e.g., speech timing), a more usable and robust system can be leveraged.

2. *Modelling users' existing work practice*: Recent research has revealed that the same students completing the same math problems experience greater load and performance deterioration as interfaces depart more from existing work practice [38, 52]. In many cases, such as mathematics education, existing work practice currently is nondigital and involves relatively 'natural' communication tools such as paper and pencil. This work practice now can be emulated in digital pen and paper interfaces as a stand-alone interface [53] or combined within a multimodal speech and pen interface [54].

3. *Supporting a broad range of representational systems*: The multimodal interface goal of supporting more powerfully expressive human–computer communication includes support for expressing a wide range of representational systems, such as linguistic, diagrammatic, symbolic and numeric, which users need for successfully completing many complex tasks (e.g., science and math). Current graphical interfaces provide only limited support for symbology and diagramming, which restricts their utility. In contrast, pen interfaces or multimodal interfaces that incorporate pen input can provide excellent coverage of different representational systems and related support for human performance [38, 52]. Multimodal interfaces should incorporate complementary input modes that offer widest coverage of representational systems.

4. *Supporting multiple linguistic codes*: Apart from supporting different modalities of communication (e.g., audio and visual) that are suitable for different user populations, multimodal interfaces also are capable of supporting a wider array of the world's languages than existing keyboard-based graphical interfaces. Current graphical interfaces are unsuitable for communicating many Asian and indigenous languages that incorporate substantial symbology (e.g.,

diacritics). As such, the present use of keyboard-based graphical interfaces for web-based interaction may be a contributing factor to rapid decline in the world's heritage of varied languages. In contrast, multimodal interfaces can be designed to support all linguistic codes.

5. *Building an interface that supports either unimodal or multimodal input*: Because users respond to dynamic changes in their own working memory limitations and cognitive load by shifting from unimodal input when load is low to multimodal input when it is high [55], an interface that can process either type of input assists users in self-managing dynamic changes in their cognitive load while solving real-world complex tasks, which in turn enhances performance.

12.5.2   Designing Low-Load Multimodal Interfaces for Education

Multimodal interfaces are considered the preferred direction for designing effective mobile interfaces, educational interfaces and interfaces for universal access, in part for reasons outlined in the previous section and elsewhere [10]. In the case of educational interface design, it is important that such interfaces be designed to minimise cognitive load, so students can focus on the intrinsic difficulty of their learning activities. Educational interfaces that effectively minimise cognitive load are more capable of supporting all students' performance, including high- and low-performing students, as well as reducing the achievement gap between student groups rather than expanding it [38].

Achieving low-load educational interface design depends on implementing many of the principles outlined here previously. For example, principle 3 (supporting a broad range of representational systems) is critical for advancing digital support of STEM education (science, technology, engineering, mathematics) because of extensive use of nonlinguistic representational systems in these important domains (e.g., numeric, symbolic, diagrammatic). In addition, principles 1 (modelling users' natural multimodal communication patterns), 3 (supporting a broad range of representational systems) and 4 (supporting multiple linguistic codes) all will be important for developing and tailoring new multimodal educational interfaces for the large world-wide population of indigenous students. Indigenous or native people on different continents historically have communicated and passed along knowledge inter-generationally using oral tradition rather than using written languages. We know that even English-speaking Caucasian students in industrialised countries experience existing graphical interfaces as higher load than either pen interfaces or nondigital tools, which compromises their performance and ability to learn [38, 52]. A relatively larger disadvantage can be expected for indigenous students using keyboard-based graphical interfaces, which would be expected to fuel a greater achievement gap between nonindigenous and indigenous groups. In the future, low-load communications and educational interfaces will need to be developed for indigenous people that focus on audio-visual and other alternative modalities rather than European/western-style keyboard input.

Recent research has investigated the design of adaptive multimodal interfaces for education. Within the domain of mathematics, an entirely implicit multimodal open-microphone engagement technique was designed that functions with high reliability (75–86%) during student group-work [56]. Basically, naturally occurring increases in students' speech amplitude and pen pressure when addressing a computer versus human interlocutor were used as cues to engage assistance from a computational tutor. This interface adapted processing to individual users, and it effectively minimised students' cognitive load by helping them remain focused on their tasks. In fact, self-reports indicated that students had limited awareness of their own adaptations.

In other research on science, a conversational multimodal interface called ISEE! science application was designed to teach marine science to elementary school children. This interface aimed to engage children as active learners and stimulate question-asking skills. It supported children's direct interaction with animated characters representing digital fish by writing and/or speaking to them. In field evaluations, 7–9-year-olds spontaneously asked an average of over 150 questions during a 1h session, underscoring the potential of multimodal interfaces for educational impact [57]. In addition, analyses of children's speech signal revealed that they adapted their acoustic signal substantially to the text-to-speech of their digital partner [55], demonstrating speech convergence with an animated character. The potential exists to design *mutually adaptive multimodal interfaces* that mimic interpersonal interactions during educational exchanges with a master teacher.

12.6   CONCLUSIONS AND FUTURE DIRECTIONS

As technologies such as handhelds, cell phones and new digital pens expand, multimodal interfaces capable of transmitting and/or processing different types of user input are proliferating, often creating new challenges. For the multimodal paradigm to achieve critical mass, the bottleneck of component technologies must still be adequately addressed: *input fusion* deserves more attention, especially in semantic level interpretation of modalities other than speech; *interaction management* will enable the next-generation of dialogue-capable interfaces. More urgently, however, evaluation methods for multimodal interfaces need to be improved beyond basic usability heuristics, taking into account the user or environmental context, with cognitive load being the obvious example. A multimodal database must also be established, so multimodal component technologies and designs can be evaluated and compared against standard benchmarks. These aspects have been fully covered in this chapter.

In many important respects, multimodal interfaces only can flourish in the future through multidisciplinary cooperation. Their successful design will continue to require guidance from cognitive science on the coordinated human perception and production of natural modalities during both interpersonal and HCI. The knowledge gained can be used to develop interface designs and recognition strategies that are much better suited to different tasks, user load levels, and environmental contexts among many others. By monitoring user's multimodal productions, it may be possible to gather implicit insights into their cognitive load status, to serve as triggers to execute adaptation strategies and alternative interaction mechanisms.

Current multimodal interfaces will require considerable further research on adaptation before they can function optimally. From a user-centred design perspective, we know that users can and do adapt more to systems than the reverse. Active adaptation technology for diverse users is necessary before achieving their full potential supporting universal access, education and other critical emerging areas. As adaptive systems become more common and increase in utility and sophistication, one long-term research agenda will be the development of mutually adaptive multimodal interfaces in which both user and system actions converge with one another – as does natural interpersonal interaction.

REFERENCES

1. M.T. Maybury, W. Wahlster (Eds.), Readings in Intelligent User Interfaces, Morgan Kaufmann Publishers, 1998.
2. R. Raisamo, Multimodal Human-Computer Interaction: a constructive and empirical study, PhD thesis, University of Tampere, Tampere, Finland, 1999.
3. P. Cohen, D. McGee, J. Clow, The efficiency of multimodal interaction for a map-based task, in: Proceedings of the sixth conference on Applied natural language processing, Morgan Kaufmann Publishers Inc, San Francisco, CA, USA, 2000, pp. 331–338.
4. S. Oviatt, Ten myths of multimodal interaction, Commun. ACM 42 (11) (1999) 74–81.
5. S. Oviatt, Multimodal interfaces for dynamic interactive maps, in: Proceedings of the ACM CHI 96 Human Factors in Computing Systems Conference, ACM, ACM, Vancouver, Canada, 1996, pp. 95–102.
6. S. Oviatt, P. Cohen, Perceptual user interfaces: multimodal interfaces that process what comes naturally, Commun. ACM 43 (3) (2000) 45–53.
7. C. Benoit, J.-C. Martin, C. Pelachaud, L. Schomaker, B. Suhm, Audio-visual and multimodal speech-based systems, in: R. Moore (Ed.), Handbook of Multimodal and Spoken Dialogue Systems: Resources, Terminology and Product Evaluation, Kluwer Academic Publishers, Boston, MA, 2000, pp. 102–203.
8. D.G. Stork, M.E. Hennecke (Eds.), Speechreading by Humans and Machines, Springer Verlag, New York, 1995.
9. G. Potamianos, C. Neti, G. Gravier, A. Garg, Automatic recognition of audiovisual speech: recent progress and challenges, in: Proceedings of the IEEE, vol. 91. IEEE, 2003, 9, pp. 1306–1326.
10. S.L. Oviatt, Multimodal interfaces, in: J. Jacko, A. Sears (Eds.), Handbook of Human-Computer Interaction: Fundamentals, Evolving Technologies and Emerging Applications (revised 2nd edition), second ed., Lawrence Erlbaum Associates, Mahwah, New Jersey, 2008, pp. 413–432.

11. L. Wu, S. Oviatt, P. Cohen, Multimodal integration – a statistical view, IEEE Trans. Multimed. 1 (4) (1999) 334–341.

12. S. Bangalore, M. Johnston, Integrating multimodal language processing with speech recognition, in: B. Yuan, T. Huang, X. Tang (Eds.), Proceedings of the International Conference on Spoken Language Processing (ICSLP2000), vol. 2, ACM, Chinese Friendship Publishers, Beijing, 2000, pp. 126–129.

13. M. Denecke, J. Yang, Partial information in multimodal dialogue, in: Proceedings of the International Conference on Multimodal Interaction (ICMI00), ACM, Beijing, China, 2000, pp. 624–633.

14. S. Kopp, P. Tepper, J. Cassell, Towards integrated microplanning of language and iconic gesture for multimodal output, in: ICMI '04: Proceedings of the 6th international conference on Multimodal interfaces, ACM, New York, NY, USA, 2004, pp. 97–104.

15. W. Wahlster, Smartkom: multimodal dialogs with mobile web users, in: Proceedings of the Cyber Assist International Symposium, Tokyo International Forum, Tokyo, 2001, pp. 33–34.

16. A.D. Baddeley, Essentials of Human Memory, Psychology Press Ltd, East Sussex, UK, 1999.

17. C.D. Wickens, Multiple resources and performance prediction, Theor. Issues Ergon. Sci. 3 (2) (2002) 159–177.

18. L.M. Encarnacao, L.J. Hettinger, Perceptual multimodal interfaces, IEEE Comput. Graph. Appl. 23 (5) (2003), pp. 24–25.

19. J. Flanagan, T. Huang, Scanning the special issue on human-computer multimodal interface, Proc. IEEE 91 (9) (2003) 1267–1271.

20. L.-P. Morency, C. Sidner, C. Lee, T. Darrell, Contextual recognition of head gestures, in: Proceedings of the Seventh International Conference on Multimodal Interfaces, ACM, ACM Press, New York, 2005, pp. 18–24.

21. M. Turk, G. Robertson, Perceptual user interfaces (introduction) Commun. ACM 43 (3) (2000) 32–34.

22. P.R. Cohen, M. Johnston, D. McGee, S. Oviatt, J. Pittman, I. Smith, et al., Quickset: multimodal interaction for simulation set-up and control, in: Proceedings of the fifth conference on Applied natural language processing, Morgan Kaufmann Publishers Inc, San Francisco, CA, USA, 1997, pp. 20–24.

23. L. Duncan, W. Brown, C. Esposito, H. Holmback, P. Xue, Enhancing virtual maintenance environments with speech understanding, Technical report, Boeing M& CT TechNet, 1999.

24. F. Chen, E. Choi, J. Epps, S. Lichman, N. Ruiz, Y. Shi, et al., A study of manual gesture-based selection for the pemmi multimodal transport management interface, in: Proceedings of the 7th International Conference on Multimodal Interfaces, ICMI 2005. ACM Press, 2005, pp. 274–281.

25. J. Bers, S. Miller, J. Makhoul, Designing conversational interfaces with multimodal interaction, in: DARPA Workshop on Broadcast News Understanding Systems, DARPA, 1998, pp. 319–321.

26. E.R. Kandell, J.H. Schwartz, T.M. Jessel, Principles of Neural Science, McGraw Hill Professional, 2000.

27. E. Smith, J. Jonides, R.A. Koeppe, Dissociating verbal and spatial working memory using PET, Cereb. Cortex 6 (1996) 11–20.

28. E. Awh, J. Jonides, E. Smith, E.H. Schumacher, R.A. Koeppe, S. Katz, Dissociation of storage and retrieval in verbal working memory: evidence from positron emission tomography, Psychol. Sci. 7 (1996) 25–31.

29. A. Kendon, Gesticulation and speech: two aspects of the process of utterances, in: M.R. Key (Ed.), Nonverbal Communication and Language, Mounton, The Hague, 1980.

30. J. Cassell, M. Stone, Living hand to mouth: psychological theories about speech and gesture in interactive dialogue systems, in: S.E. Brennan, A. Giboin, D. Traum (Eds.), Working Papers of the AAAI Fall Symposium on Psychological Models of Communication in Collaborative Systems, Menlo Park, California, 1999, pp. 34–42. American Association for Artificial Intelligence.

31. G. Miller, The magical number, seven plus or minus two: some limits on our capacity of processing information, Psychol. Rev. 63 (1954) 81–97.

32. L. Peterson, M. Peterson, Short-term retention of individual verbal items, 58 (1959) 193–198.

33. F. Paas, A. Renkl, J. Sweller, Cognitive load theory: instructional implications of the interaction between information structures and cognitive architecture, Instr. Sci. 32 (2004) 1–8.

34. F. Paas, J.E. Tuovinen, H. Tabbers, P.W.M. Van Gerven, Cognitive load measurement as a means to advance cognitive load theory, Educ. Psychol. 38 (1) (2003) 63–71.

35. S.Y. Mousavi, R. Low, J. Sweller, Measurement and analysis methods of heart rate and respiration for use in applied environments, J. Educ. Psychol. 87 (2) (1995) 319–334.

36. S. Oviatt, R. Coulston, R. Lunsford, When do we interact multimodally?: Cognitive load and multimodal communication patterns, in: ICMI '04: Proceedings of the 6th international conference on Multimodal interfaces, ACM, New York, NY, USA, 2004, pp. 129–136.

37. S. Goldin-Meadow, H. Nusbaum, S. Kelly, S. Wagner, Explaining math: gesturing lightens the load, Psychol. Sci. 12 (2001) 516–522.

38. S. Oviatt, Human-centered design meets cognitive load theory: designing interfaces that help people think, in: MULTIMEDIA '06: Proceedings of the 14th

annual ACM international conference on Multimedia, ACM, New York, NY, USA, 2006, pp. 871–880.

39. M.W. Alibali, S. Kita, A.J. Young, Gesture and the process of speech production: we think, therefore we gesture, Lang. Cogn. Process. 15 (6) (2000) 593–613.

40. S.L. Oviatt, Multimodal interactive maps: designing for human performance, Hum. Comput. Interact. 12 (1997) 93–129.

41. R.H. Logie, Visuo-Spatial Working Memory, Lawrence Erlbaum Associates, Hove, UK, 1995.

42. A.S. Bregman, Auditory Scene Analysis, MIT Press, Cambridge, MA, 1990.

43. S. Oviatt, R. Coulston, S. Tomko, B. Xiao, R. Lunsford, M. Wesson, et al. Toward a theory of organized multimodal integration patterns during human-computer interaction, in: ICMI '03: Proceedings of the 5th international conference on Multimodal interfaces, ACM, New York, NY, USA, 2003, pp. 44–51.

44. K. Koffka, Principles of Gestalt Psychology, Harcourt, Brace and Company, NY, 1935.

45. A. Kendon, Do gestures communicate? a review, Res. Lang. Soc. Interact. 27 (3) (1994) 175–200.

46. S. Oviatt, A. DeAngeli, K. Kuhn, Integration and synchronization of input modes during multimodal human-computer interaction, in: S. Pemberton (Ed.), SIGCHI conference on Human Factors in computing systems, Atlanta, Georgia, United States, 1997, ACM, pp. 415–422.

47. N. Ruiz, Measuring Cognitive Load in Multimodal Interfaces, PhD thesis, UNSW, Sydney, Australia, 2008.

48. N. Ruiz, R. Taib, F. Chen, Examining the redundancy of multimodal input, in: J. Kjeldskov, J. Paay (Eds.), Proc. 20th annual conference of the Australian computer-human interaction special interest group (OzCHI'06), Sydney, Australia, 2006, pp. 389–392.

49. C. Müller, B. Grossmann-Hutter, A. Jameson, R. Rummer, F. Wittig, Recognising time pressure and cognitive load on the basis of speech: an experimental study, in: J. Vassileva, P. Gmytrasiewicz, M. Bauer (Eds.), UM2001, User Modeling: Proceedings of the Eighth International Conference, Springer, Berlin, 2001, pp. 24–33.

50. M.A. Khawaja, N. Ruiz, F. Chen, Potential speech features for cognitive load measurement, in: OZCHI '07: Proceedings of the 2007 conference of the computer-human interaction special interest group (CHISIG) of Australia on Computer-human interaction: design: activities, artifacts and environments, ACM, New York, NY, USA, 2007, pp. 57–60.

51. B. Yin, F. Chen, Towards automatic cognitive load measurement from speech analysis, in: J. Jacko (Ed.), Human-Computer Interaction. Interaction Design and Usability, vol. 4550, Springer Berlin, Heidelberg, 2007, pp. 1011–1020.

52. S.L. Oviatt, A. Arthur, Y. Brock, J. Cohen, Expressive pen-based interfaces for math education, in: C. Chinn, G. Erkerns, S. Puntambekar (Eds.), Proceedings of the Conference on Computer Supported Collaborative Learning 2007: Of Mice, Minds and Society, International Society of the Learning Sciences, 2007, pp. 573–582.

53. Anoto, About anoto, March 2009. <http://www.anoto.com/about-anoto-1.aspx>. Accessed 5 September 2009.

54. Livescribe, Homepage, 2009. <http://www.livescribe.com/>. Accessed 5 September 2009.

55. S. Oviatt, C. Darves, R. Coulston, Toward adaptive conversational interfaces: modeling speech convergence with animated personas, in: Transactions on Human Computer Interaction (TOCHI), vol. 11, 2004, pp. 300–328 (Chapter 3) ACM.

56. S. Oviatt, C. Swindells, A. Arthur, Implicit user-adaptive system engagement in speech and pen interfaces, in: Conference on Human Factors in Computing Systems (CHI08), CHI Letters, ACM, ACM Press, Crete, Greece, 2008, pp. 969–978.

57. C. Darves, S. Oviatt, Talking to digital fish: designing effective conversational interfaces for educational software, in: Z. Ruttkay, C. Pelachaud (Eds.), From Brows to Trust: Evaluating Embodied Conversational Agents, vol. 7, Kluwer, Dordrecht, 2004, pp. 271–292.

Chapter 13

Multimodal HCI Output: Facial Motion, Gestures and Synthesised Speech Synchronisation

Igor S. Pandžić
University of Zagreb, Croatia

13.1 INTRODUCTION

In this chapter, we present an overview of the issues involved in generating multimodal output consisting of speech, facial motion and

Multimodal Signal Processing, ISBN: 9780123748256

gestures. We start by introducing a basic audio-visual (AV) speech synthesis system that generates simple lip motion from input text using a text-to-speech (TTS, see Chapter 3) engine and an animation system. Throughout this chapter, we gradually extend and improve this system first with coarticulation, then full facial motion and gestures and finally we present it in the context of a full embodied conversational agent (ECA) system. At each level, we present key concepts and discuss existing systems.

We concentrate on real-time interactive systems, as necessary for human–computer interaction (HCI). This requires on-the-fly generation of speech and animation and their synchronisation and does not allow for any time-consuming preprocessing. We discuss the practical issues that this requirement brings in the final section that deals with obtaining timing information from the TTS engine.

We concentrate on systems that deal with plain text input (ASCII or UNICODE) rather than those that require manual tagging of text because such systems add a significant overhead to the implementation of any HCI application.

13.2 BASIC AV SPEECH SYNTHESIS

The basic AV speech synthesis system, as shown in Figure 13.1 (see Chapter 3 for more details on speech synthesis), is fairly simple to understand and build. It can provide satisfactory results for many applications that do not require fully accurate lip motion, especially if used with nonrealistic virtual characters such as cartoon figures. It is also a good starting point for understanding the more accurate and complete systems described later in this chapter.

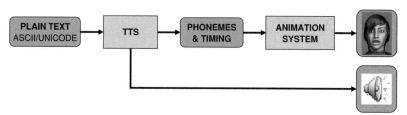

FIGURE 13.1   Basic audio-visual speech synthesis system.

The TTS system is at the core of any AV speech synthesis system. TTS generates and outputs the speech sound based on plain text input in ASCII or UNICODE encoding. Crucially for AV speech synthesis, TTS provides information about the phonemes it generates and the time of their occurrence. Normally, any TTS system can provide such information; though they vary in the implementation of the mechanism for passing this information to other components. Common mechanisms include a callback function to be implemented by the information user and called by the TTS at appropriate times or a fast preprocessing step generating a list of phonemes and other events. A more detailed discussion about these mechanisms is provided in the section on TTS timing issues.

For now, we can assume that the animation system receives the information about each phoneme at the latest by the time the phoneme gets pronounced and that there is a common time base between the TTS and the animation system. Based on this, the animation system can simply map each phoneme to a mouth shape and display that mouth shape at the appropriate time, i.e., while the corresponding phoneme is pronounced.

It is important to note that several phonemes may map to the same mouth shape. The mouth shape corresponding to the pronunciation of one or more phonemes is called a viseme [1]. It is very common to use a fairly small set of visemes (typically between 7 and 15) to move the mouth. A standardised set of visemes exists in the MPEG-4 International Standard [2], within the Face and Body Animation (FBA) specification [3]. The MPEG-4 set of visemes is presented in Table 13.1.

Simply mapping each phoneme to a viseme, displaying the viseme and then switching abruptly to another viseme when TTS generates the new phoneme would result in very jerky animation. Therefore, all systems deploy at least a simple interpolation scheme to smooth the animation. The simplest scheme performs a linear interpolation between the existing viseme and the new viseme over a fixed period called the onset time (e.g., 100 ms). The animation system performs a blend (morph) between the two mouth shapes at each time frame, corresponding to the interpolated weights of the two visemes. Thus, there is a smooth transition from one mouth shape to another

TABLE 13.1 MPEG-4 visemes and related phonemes

Viseme #	Phonemes	Example	Viseme #	Phonemes	Example
0	none	na	8	n, l	lot, not
1	p, b, m	put, bed, mill	9	r	red
2	f, v	far, voice	10	A:	car
3	T,D	think, that	11	e	bed
4	t, d	tip, doll	12	I	tip
5	k, g	call, gas	13	Q	top
6	tS, dZ, S	chair, join, she	14	U	book
7	s, z	sir, zeal			

during the onset time. A slightly more elaborate scheme assumes that phoneme/viseme timing is available in advance. This makes it possible to calculate the duration of appearance for each viseme and adjust the transition periods proportionally to this duration.

Although these simple schemes provide acceptable results for many applications, they do not produce realistic mouth motion. For this, a coarticulation scheme is needed. Before discussing coarticulation, we need to briefly introduce the animation system.

13.3 THE ANIMATION SYSTEM

Animation of human-like characters is a very large topic and a detailed presentation would be out of the scope of this book. In this section, we provide a brief overview with the emphasis on control issues that are important for understanding how the animation system connects with other components to perform AV speech synthesis.

For the purpose of presentation, we introduce three levels of character animation control: behavioural control, motor control and geometrical control. Figure 13.2 presents the conceptual view of the animation system with emphasis on these control levels. Although many practical implementations blur the boundaries between the conceptual modules outlined in the figure or even skip some of the modules, this view is useful for understanding the issues in character animation control and implementation.

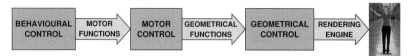

FIGURE 13.2   A conceptual view of the character animation system.

At the level of *behavioural control*, we deal with fairly high-level control of the character, controlling gestures, expressions, gaze, locomotion and speech. At this level, the character is often controlled by a high-level markup language such as behaviour markup language [4, 5] that includes mechanisms for specifying synchronisation between various behavioural elements. Most importantly in the current context, such a markup language typically allows specification of synchronisation points between speech and animation. This is done by inserting tags within the text to be spoken by the character. How these tags may be used to actually produce synchronised speech and animation is covered later in this chapter.

The *motor functions* translate the higher level behavioural controls into the lower level motor controls. The main approaches for doing this are

1. using libraries of predefined animation sequences produced either by artists or using motion capture techniques,
2. procedural approaches, usually involving forward and inverse kinematics, and
3. hybrid approaches.

Using predefined sequences can result in high-quality animation, but control is severely limited to the set of available sequences. Conversely, procedural approaches, such as [6], offer great flexibility of control but at the expense of naturalness of motion. Hybrid approaches rely on libraries of high-quality predefined animation sequences but include procedures to generate new contents by modifying and/or combining the existing materials. Parametric motion graphs [7] are a good example of such an approach.

At the level of *motor control*, we deal with control parameters that directly move physical parts of the body and face. For body motion, the usual approach is to control the joints of a simplified skeleton that moves the body. Configurations of the skeleton vary in complexity. MPEG-4 FBA [2] specifies a rather complex skeleton with 75 degrees of freedom (DOF) for the body and additional 30 DOF for each hand. In practice, only a subset of these DOFs is typically used. For the face, perhaps the strongest influence on the parameterisation was that of the Facial Action Coding System (FACS) [8]. FACS was not meant for animation; it is rather a system for evaluating (scoring) the facial expressions on the faces of real humans to code them in a standardised way comprising a set of 46 Action Units or basic facial movements. However, due to its systematic approach, foundation on muscle activity and detailed description of visual effects of each Action Unit, FACS had obvious attractions to the researchers implementing FA systems. Several systems, in particular muscle-based ones, used subsets of FACS as control parameters. This and several other systems influenced the standardisation process that resulted in the MPEG-4 FBA specification [3]. It specifies 66 low-level parameters that move various parts of the face and also specifies 15 visemes and 6 basic expressions. A particular problem faced at this level is the combination of conflicting facial movements, e.g., when a facial expression such as a smile requires motor control parameters that conflict with those currently needed for speech [9].

The *geometrical functions* translate the motor control parameters into *geometrical controls*, i.e., actual movements of the 3D model geometry – usually the vertices of a 3D polygon mesh and/or transform nodes in the 3D scene graph. The most common approach to do this for the body is skinning (sometimes also known as bone-based animation). The movement of each vertex is controlled by weighted influence of one or more bones. An overview of this and related methods is available in [10]. A common approach for controlling the face is morphing. This involves defining a number of key positions of the face by displacing its vertices. These key positions are called morph targets. The animation system then interpolates the vertex positions between the neutral model and one or more morph targets, each with an assigned weight factor. Traditionally, morph targets are used for high-level facial motions like visemes or expressions. However, they have been successfully extended to low-level facial motions, in particular the MPEG-4 facial animation parameters [11]. Another approach is to use skinning for facial animation, with the advantage of using the same technique for both body and face. Although the bone rig used for facial animation has almost nothing in common with the actual human skull, the method gives very good results. There are numerous other approaches ranging from ones purely based on observation to pseudo muscle models and various degrees of physical simulation of bone, muscle and tissue dynamics.

Finally, the *rendering engine* draws the animated character on the screen. It should be noted that current rendering engines typically include the above-mentioned skinning and morphing functions and thus may encapsulate the geometrical functions. Some popular examples are open source engines Ogre, Horde3D and Panda3D[1], though there are numerous others.

For the purposes of AV speech synthesis, the animation is controlled at either motor or behavioural level.

13.4 COARTICULATION

The previously introduced simple AV synthesis makes an implicit assumption that speech is a sequence of discrete segments (phonemes) with identifiable boundaries between them. Unfortunately, this assumption is wrong. Articulatory movements overlap in a very complex fashion and may be influenced by up to five segments before or after the current one [12]. The practical problem lies in the fact that TTS systems typically generate a series of discrete phoneme occurrences, so this is the only input available for the visual synthesis. Coarticulation models attempt to reproduce the correct overlapping articulatory movements based on this input. A correct coarticulation is crucial for realistic visual speech synthesis. When implemented correctly, the resulting AV speech synthesis is realistic and accurate enough to improve speech intelligibility in noisy conditions, even when the character model itself does not look realistic [13].

Regardless of which coarticulation model is used, it is important to notice that the animation system needs to be controlled by lip shape parameters rather than visemes. Research has shown that only

---

1. http://www.ogre3d.org/, http://www.horde3d.org/ and http://panda3d.org/.

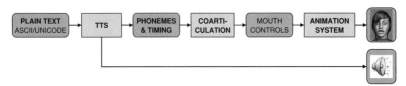

FIGURE 13.3   Audio-visual speech synthesis system with coarticulation.

a few parameters are necessary to describe the mouth shapes. These parameters are lip height (LH), lip width (LW), upper-lip protrusion and lower-lip protrusion [14]. These parameters can be mapped to motor control level of an animation system, e.g., to the MPEG-4 facial animation parameters [15]. Figure 13.3 shows the AV speech synthesis system that includes coarticulation.

The most widely used coarticulation model is the one introduced by Cohen and Massaro [16]. It is based on Löfqvist's theory of speech production [17]. This model uses overlapping dominance functions to model the influence of each viseme on facial control parameters. The facial control parameters used in this model are LH, LW, lip protrusion and jaw rotation (slightly different from the ones mentioned earlier in this section). The dominance functions have the form of negative exponential functions centred within one speech segment and scaled with a magnitude factor characteristic for the viseme produced in this speech segment. They are generated for each control parameter separately. They overlap in time and thus produce the final curve of each control parameter. Several models improve the Cohen and Massaro approach or take different approaches to the coarticulation problem. A good overview of these models is available in [15].

13.5   EXTENDED AV SPEECH SYNTHESIS

Techniques described so far in this chapter deal only with lip motion. A character standing completely still, with frozen face and only the lips moving, would look extremely unnatural. For virtual characters to appear believable, it is necessary to generate not just lip motion but a full set of accompanying actions, potentially all of the following: head movements (nods, swings, shakes, tilts and so on), gaze, eye blinks, eyebrow movements (raising, frown), expressions, body shifts and gestures.

The simplest solution is to introduce predefined background motion, e.g., a motion capture sequence driving the head and body posture, eye blinks, eyebrows and gaze. The sequence must be fairly long in order not to be repetitive. Another solution is to apply random functions to these animation elements. Perlin noise [18] has been successfully used for this purpose [19, 20]. Both approaches provide a significant improvement over a static talking head, but the accompanying actions are not in any way related to speech. This produces a severely unrealistic result. When real people speak, the accompanying actions are highly related to speech at all levels. On the prosodic level, stressed parts of speech may coincide with actions such as nods and eyebrow movements. On the lexical level, introducing new terms into the discourse may trigger particular gestures. On the semantic level, the nonverbal actions represent a communication channel that may be even more important than the verbal channel [21, 22].

Automatically generating believable accompanying actions in relation to speech is a subject of ongoing research. The ideas generally fall into two main categories. In the rule-based approach, researchers study the theory of conversation and discourse and try to build theoretically grounded rules to drive accompanying actions. The data-driven approaches rely on machine learning techniques to recreate the behaviour based on recorded examples. Both approaches require similar additions to the system architecture. Figure 13.4

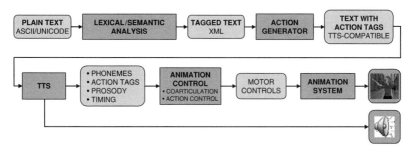

FIGURE 13.4 Concept of a complete audio-visual speech synthesis system.

presents a concept of a complete AV speech synthesis system architecture incorporating ideas from various previous works. We use this conceptual architecture to illustrate how a complete system may work, before proceeding with an overview of existing systems.

Our conceptual system takes plain text as input and generates synthetic speech with coarticulation and a full set of accompanying facial and body actions, in real time and with minimal delay. The purpose of lexical/semantic analysis module is to extract as much information as possible from the text. This information may include semantic tags [23], theme/rheme classification [24], punctuation and new term occurrences [25] and other information. Typically the lexical/semantic analysis module embeds the extracted information into the text in form of tags, usually in a form of XML. The information is then used by the action generator module to generate appropriate actions. This is the core of the system and implementations differ widely. The actual approaches are discussed in the next subsections. The action generator would typically insert some form of action tags into the text. The form must be compatible with the TTS system used, e.g., for Microsoft Speech Application Programming Interface (SAPI) systems the tags may be inserted in form of the SAPI bookmark tag. The TTS system generates speech and provides the necessary information to the animation control module. Beside the previously discussed timed phonemes (or visemes – some TTS systems, such as SAPI, may directly provide visemes), TTS must also provide the time of occurrence for the action tags inserted into text. These times serve as synchronisation points for the actions. The TTS system may also provide prosodic information, which can be very useful in generating accompanying actions, though this requires a fairly tight integration with the TTS, and we have found only one system that actually does this [26]. Other than coarticulation, the extended animation control has the task of generating the accompanying actions at appropriate times and controlling them in terms of motor control parameters used by the animation system. It must integrate coarticulation and other actions into a single stream of control parameters that drives the animation system. The animation system produces and displays the final animation.

13.5.1   Data-Driven Approaches

The data-driven approaches rely on a training set of data collected from real speakers by means of video recording, manual annotation, motion capture or other. In general, the training set consists of the behaviour data and the input data. The behaviour data may include gesture/action annotations, detailed motion of the head or other body parts, eye movements, facial feature tracking, etc. The main point is that this data is an as-accurate-as-possible record of the behaviour of real people in real-life situations (sometimes acted behaviour is used due to practical issues in capturing real-life situations). The input data can be any data that is correlated with the behaviour data. Examples include occurrences of new terms in the text, punctuation marks, semantic tags obtained by text analysis, prosodic information, etc. In the context of data-driven AV speech synthesis, it is assumed that the input data 'drives' the behaviour, even though this may not be the case in real human behaviour (and indeed, usually it is not). Some form of supervised learning is then applied to create a computational model that can generate the behaviour data from input data. The computational model can then generate new behaviour data from previously unseen samples of input data. If the training set was accurate and the model well-built, the generated behaviour will retain the quality of the real behaviour while not being exactly the same. In the following paragraphs, we briefly introduce four specific examples of this approach.

Wang et al. [26] introduce an AV speech synthesis system capable of generating head motions, eye blinks, gaze and hand gestures. This is a rare example where the same team develops both the speech synthesis system and the animation system. This allows a very tight integration of both systems. The information provided by the TTS is far richer than usual and consists of semantic and prosody features including phonetic parameters, timing, part-of-speech, word accent and stress, pitch contour and energy contour, all of which may be used to drive the animations. Specifically, head motion is generated using a Hidden Markov Model-based motion generation method adapted from [27, 28]. A head tracker is used to obtain head motion in real videos (behaviour data), whereas the prosody features (F0 and energy) are used as input data.

The *Eyes Alive* system [29] implements an eye movement model based on empirical models of saccades (rapid eye movements) and statistical models of eye-tracking data. The model reproduces eye movements that are dynamically correct at the level of each movement and that are also globally statistically correct in terms of the frequency of movements, intervals between them and their amplitudes. Although speaking and listening modes are distinguished, movements are not related to the underlying speech contents, punctuation, accents, etc. The system is not originally designed to work in the speech synthesis context, but the collected statistical data and knowledge on eye movements may be applied with certain modifications to the speech synthesis scenario.

Neff et al. [23] introduce a data-driven approach to generate hand gestures. The system can capture and reproduce the personal style of a speaker. The behaviour data is annotated manually in training videos and consists of a lexicon of gestures. The input data is abstracted from the transcript text. The text is first manually tagged with theme, rheme and focus, then automatically processed to obtain semantic tags such as AGREEMENT ('yes'), PROCESS ('create', 'manage') and PERS_NAME ('Michael Jackson'). There are 87 semantic tags, and they are used as input data for the algorithm. In the runtime phase, the system still requires manual annotation of input text with theme, rheme and focus, but methods exist to obtain such information automatically [24].

Smid et al. [30] propose a universal architecture for statistically based HUman GEsturing. It is used for producing and using statistical models for facial gestures based on any kind of input data – it can be any kind of signal that occurs in parallel to the production of gestures in human behaviour and that may have a statistical correlation with the occurrence of gestures. Examples of possible input data include text that is spoken, audio signal of speech, bio signals, etc. The correlation between the input data and the gestures is used to first build the statistical model of gestures based on a training corpus consisting of sequences of gestures and corresponding input data sequences. In the runtime phase, the raw, previously unknown input data, is used to trigger the real-time gestures of the agent based on the previously constructed statistical model. This universal architecture is useful for experimenting with various kinds of potential input signals and their features, as well as exploring the correlation of such signals or features with the gesturing behaviour.

13.5.2   Rule-Based Approaches

The majority of researchers attempting to generate full AV speech base their work on studies of human behaviour related to speech and conversation. Detailed studies exist for many specific aspects of speech-accompanying actions at different levels, e.g., relation between speech and eyebrow motion [31, 32, 33, 34]; relation between voice parameters such as amplitude and fundamental frequency and head motion [35, 36]; influence of eyebrow and head motion on the perception of prominence [37, 38] and the use of gaze [39, 40]. Rules derived from such studies are codified and used to drive the character behaviour.

The *BEAT system* [24] controls movements of hands, arms and the face and the intonation of the voice, relying on rules derived from the extensive research in the human conversational behaviour. It takes plain text as input and annotates it with contextual and linguistic information tags including clause, theme and rheme, word newness, contrast, and objects and actions. BEAT applies a set of rules to these language tags to generate actions such as beat gestures, iconic gestures, raising eyebrows to signal introduction of new material, gaze away/toward listener for theme/rheme, etc.

Pelechaud et al. in [41] report results from a program that produces animation of facial expressions and head movements, conveying information correlated with the intonation of the voice. Facial expressions are divided by their function into conversational signals, punctuators, manipulators and regulators. Only functions related directly to pattern of voice are included. A set of rules generates appropriate facial actions for each of these functions. Actions include head and eye movements. The input to the program is a file containing an utterance already decomposed and written in its phonological representation with its accents marked.

Lundeberg et al. in [42] developed a talking head with the purpose of acting as an interactive agent in a dialogue system. To add to the realism and believability of the dialogue system, emotional cues, turn-taking signals and prosodic cues such as punctuators and emphasisers were given to the agent. The main rules for creating the prosodic gestures were to use a combination of head movements and eyebrow motion and maintain a high level of variation between different utterances.

Lee et al. in [43] have created rules for generating nonverbal behaviour by extracting information from the lexical, syntactic and semantic structure of the surface text. Behaviour includes head movement, eyebrow movement, gaze/eye movement, shoulder shrug and mouth pulled on one side. Each rule has associated nonverbal behaviours and a set of words that are usually spoken with it. Examples include contrast – head moved to the side and brow raise co-occurring with words *but, however*; interjection – head nod, shake or tilt co-occurring with *yes, no, well*; inclusivity – lateral head sweep co-occurring on *everything, all, whole, several, plenty, full*; etc. Affect state and emphasis additionally influence the rules for generating nonverbal behaviour.

13.6   EMBODIED CONVERSATIONAL AGENTS

So far we have viewed the AV speech synthesis system mainly in isolation. In this section, we put it into the full context of an ECA system. ECAs [44] are graphically embodied virtual characters that can engage in a meaningful conversation with human user(s). Ultimately, this conversation should be fully multimodal, involving verbal and nonverbal channels in both directions just like the conversation between humans. In this sense, we can view ECAs as the ultimate multimodal HCI systems.

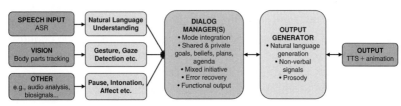

FIGURE 13.5   Concept of a full embodied conversational agent system.

Figure 13.5 presents a conceptual view of an ECA architecture allowing such conversation to proceed. The concept does not present any particular system but is based on recent trends in ECA development [5, 45]. The architecture consists of input modules on the left, output modules on the right and dialogue management in the centre. Both input and output sides are layered. On the input side, lower level inputs – such as plain text from ASR, raw body part movements from tracking, etc. – are analysed and translated into higher level concepts such as dialogue acts, gestures, gaze, intonation, pause, affect, etc. These high level chunks of information from different modalities are then integrated in the dialogue management unit and used to produce equally high-level functional outputs. It is the output generator that translates these functional outputs into language, nonverbal signals and prosody, which are fed to the output module. The output module may be the whole AV speech synthesis system similar to the one presented in Figure 13.4. However, in this context, the animation controls, on the behavioural level, are already generated by the output generator, so the output module itself is simplified and works mainly on the motor level. The backward arrows from output and output generator modules indicate the possibility of feedback, e.g., about accomplishment of task or about the impossibility to perform certain actions.

In this kind of context text, nonverbal signals and prosody are all generated by the same unit and driven by a higher level behaviour agenda.

Pelechaud et al. in [46] give ECA some aspects of nonverbal communication using a taxonomy of communicative behaviour as proposed by Poggi [40]: information on the speaker's belief, intention, affective and meta-cognitive state. Each of those functions generates signals such as facial expressions, gaze behaviour and head movement. The system takes a text and a set of communication functions as input.

Cassel et al. in [47] automatically generate and animate conversations between multiple human-like agents. Conversations are created by a dialogue planner that produces the text as well as the intonation of the utterances, which then together with the speaker/listener relationship drive hand gestures, lip motions, eye gaze, head motion (nod) and facial expressions – conversational signals, punctuators, manipulators and emblems following principles of synchrony. A conversational signal starts and ends with the accented word, punctuator happens on the pause, blink is synchronised at the phoneme level, whereas emblems are performed consciously and must be specified by the user.

Kenny et al. in [48] describe probably the most comprehensive ECA system built to date, the Virtual Human Architecture. It is a distributed, multilayered architecture that integrates a large number of components needed to create a virtual human in a cognitively and psychologically plausible way. At the cognitive layer, the intelligent agent module is reasoning about which actions to take, generating communication acts and performing emotional modelling. The virtual human layer consists of modules that perform speech recognition and synthesis, natural language understanding and generation, nonverbal behaviour generation based on [43] and the SmartBody interactive character animation system. Finally, the simulation layer consists of a graphics engine, interface modules for various gaze, gesture and face tracking systems that have been connected to the architecture.

13.7 TTS TIMING ISSUES

In this section, we deal with the issue of obtaining timing information from the TTS system. This issue is common to all implementations of AV speech. In general, any TTS system is capable of providing the timing information for the occurrence of phonemes, words and tags (bookmarks) inserted into the text. The issue is when this information is available. If the information is available in advance, before the utterance starts, it is possible to perform a priori planning and synchronisation. However, if it becomes available only at the time of occurrence, a priori planning is not possible, and we are limited to on-the-fly synchronisation.

13.7.1 On-the-Fly Synchronisation

The limits of on-the-fly synchronisation are twofold and relate to coarticulation and gesture synchronisation. For coarticulation, the problem lies in the fact that next phonemes are not available in advance, and it is therefore not possible to take them into account in the coarticulation model. This practically means that none of the existing coarticulation models can be fully implemented. It is possible to adapt the coarticulation model to perform only backward coarticulation, relying only on the past information, but the results will obviously be inferior to the full model.

The problem with gesture synchronisation is slightly less obvious. It occurs in the situations when it is necessary to synchronise a speech event with a particular phase of a gesture. Consider an example of a beat gesture with the duration of 800 ms, with the stroke occurring at 400 ms. If we need to synchronise the stroke with the beginning of a word, the gesture should start 400 ms before the beginning of the word. In the on-the-fly synchronisation framework, this is clearly impossible – by the time the information about the word is available, it is already too late to start the gesture.

Unfortunately this is currently the case with the most prominent commercial TTS platform, the SAPI interface [49]. SAPI is an event-based framework. The TTS engine generates events (phonemes, bookmarks, etc.) as they occur. The application that wants to synchronise to the TTS engine must catch these events and react to them. There is no possibility to obtain the timing in advance through a preprocessing step (other than running the whole utterance as a preprocessing step, but this generates delay that is unacceptable for real-time applications).

13.7.2   A Priori Synchronisation

A priori synchronisation assumes that the timing information for the whole utterance is available before the utterance starts. This allows for full coarticulation and animation planning. The problem lies in obtaining the timing in advance.

Some TTS systems (e.g., [50, 51]) can perform a fast preprocessing step and provide the full timing information with an acceptable delay. Many AV speech synthesis systems use such TTS engines [52, 53, 24]. However, this approach severely limits the choice of TTS. Most commercial TTS engines use the SAPI standard, which does not offer the timing information in advance.

There have been attempts to use the preprocessing step of one TTS engine and apply the obtained timing to another TTS engine; however, no results have been published yet. A promising approach is to use a machine learning technique to predict the timing. This has been done in [54] using neural networks with acceptable results for gesture synchronisation, though not for coarticulation.

13.8 CONCLUSION

This chapter provides a structured overview of issues involved in generating multimodal output consisting of speech, facial motion and gestures for the purpose of HCI. We concentrated on systems dealing with plain text input and automatically generating full multimodal output in real time, suitable for integration into larger HCI systems. For the purpose of presentation, we have first outlined the basic AV speech synthesis system and then gradually extended with more complex features. At each level, we have presented key concepts and discussed existing systems.

REFERENCES

1. C. Fisher, Confusions among visually perceived consonants, J. Speech Hear. Res. 15 (1968) 474–48.
2. ISO/IEC 14496 – MPEG-4 International Standard, Moving Picture Experts Group.
3. I.S. Pandzic, R. Forchheimer (Eds.), MPEG-4 Facial Animation: The Standard, Implementation and Applications, John Wiley & Sons, Inc., New York, NY, USA, 2002.
4. S. Kopp, B. Krenn, S. Marsella, A. Marshall, C. Pelachaud, H. Pirker, et al., Towards a common framework for multimodal generation: the behavior markup language, in: Intelligent Virtual Agents, 2006, pp. 205–217.
5. H. Vilhjálmsson, N. Cantelmo, J. Cassell, N.E. Chafai, M. Kipp, S. Kopp, et al., The behavior markup language: recent developments and challenges, in: IVA'07: Proceedings of the 7th International Conference on Intelligent Virtual Agents, Springer-Verlag, 2007, pp. 99–111.
6. S. Kopp, I. Wachsmuth, Model-based animation of coverbal gesture, in: Proceedings of Computer Animation, IEEE Press, 2002, pp. 252–257.

7. R. Heck, M. Gleicher, Parametric motion graphs, in: I3D'07: Proceedings of the 2007 Symposium on Interactive 3D Graphics and Games, ACM, New York, NY, USA, 2007, pp. 129–136.

8. P. Ekman, W. Friesen, Manual for the Facial Action Coding System, Consulting Psychologists Press, Inc., Palo Alto, CA, 1978.

9. T.D. Bui, D. Heylen, A. Nijholt, Combination of facial movements on a 3D talking head, in: CGI'04: Proceedings of the Computer Graphics International, IEEE Computer Society, Washington, DC, USA, 2004, pp. 284–291.

10. A. Mohr, M. Gleicher, Building efficient, accurate character skins from examples, in: SIGGRAPH'03: ACM SIGGRAPH 2003 Papers, ACM, New York, NY, USA, 2003, pp. 562–568.

11. I.S. Pandzic, Facial motion cloning, Graph. Models 65 (6) (2003) 385–404.

12. R. Kent, F. Minifie, Coarticulation in recent speech production models, J. Phon. 5 (1977) 115–133.

13. I.S. Pandzic, J. Ostermann, D. Millen, User evaluation: synthetic talking faces for interactive services, The Vis. Comput. J. 15 (7–8) (1999) 330–340.

14. C. Magno-Caldognetto, E. Zmarich, P. Cosi, Statistical definition of visual information for Italian vowels and consonants, in: D. Burnham, J. Robert-Ribes, E. Vatikiotis-Bateson (Eds.), International Conference on Auditory-Visual Speech Processing AVSP'98, 1998, pp. 135–140.

15. C. Pelachaud, Visual text-to-speech, in: I. Pandzic, R. Forchheimer (Eds.), MPEG-4 Facial Animation – The standard, implementations and applications, John Wiley & Sons, 2002, pp. 125–140.

16. M. Cohen, D. Massaro, Modeling coarticulation in synthetic visual speech, in: V. Magnenat-Thalmann, D. Thalmann (Eds.), Models and Techniques in Computer Animation, Springer-Verlag, Tokyo, 1993, pp. 139–156.

17. A. Löfqvist, Speech as audible gestures, in: W. Hardcastle, A. Marchal (Eds.), Speech Production and Speech Modeling, NATO ASI Series, vol. 55, Kluwer Academic Publishers, Dordrecht, 1990, pp. 289–322.

18. K. Perlin, Improving noise, in: SIGGRAPH '02: Proceedings of the 29th Annual Conference on Computer Graphics and Interactive Techniques, ACM, New York, NY, USA, 2002, pp. 681–682.

19. K. Perlin, A. Goldberg, Improv: a system for scripting interactive actors in virtual worlds, in: SIGGRAPH '96: Proceedings of the 23rd Annual Conference on Computer Graphics and Interactive Techniques, ACM, New York, NY, USA, 1996, pp. 205–216.

20. K. Perlin, Layered compositing of facial expression, in: SIGGRAPH'97 Technical Sketch, New York University Media Research Lab. ≪http://mrl.nyu.edu/projects/improv/sig97-sketch/, http://mrl.nyu.edu/perlin/demox/Face.html≫, 1997.

21. R. Birdwhistell, Kinesics and Context, University of Pennsylvania Press, Philadelphia, 1970.
22. A. Mehrabian, Silent Messages, Wadsworth, Belmont, CA, 1971.
23. M. Neff, M. Kipp, I. Albrecht, H.-P. Seidel, Gesture modeling and animation based on a probabilistic re-creation of speaker style, ACM Trans. Graph. 27 (1) (2008) 1–24.
24. J. Cassell, H.H. Vilhjálmsson, T. Bickmore, Beat: the behavior expression animation toolkit, in: SIGGRAPH '01: Proceedings of the 28th Annual Conference on Computer Graphics and Interactive Techniques, ACM, 2001, pp. 477–486.
25. K. Smid, I. Pandzic, V. Radman, Autonomous speaker agent, in: Computer Animation and Social Agents Conference CASA 2004, 2004.
26. L. Wang, X. Qian, L. Ma, Y. Qian, Y. Chen, F.K. Soong, A real-time text to audio-visual speech synthesis system, in: Proceedings of Interspeech, 2008.
27. Z. Deng, S. Narayanan, C. Busso, U. Neumann, Audio-based head motion synthesis for avatar-based telepresence systems, in: ETP'04: Proceedings of the 2004 ACM SIGMM Workshop on Effective Telepresence, ACM, New York, NY, USA, 2004, pp. 24–30.
28. G. Hofer, H. Shimodaira, J. Yamagishi, Speech driven head motion synthesis based on a trajectory model, in: SIGGRAPH'07: ACM SIGGRAPH 2007 Posters, ACM, New York, NY, USA, 2007, pp. 86.
29. S.P. Lee, J.B. Badler, N.I. Badler, Eyes alive, in: SIGGRAPH '02: Proceedings of the 29th Annual Conference on Computer Graphics and Interactive Techniques, ACM, New York, NY, USA, 2002, pp. 637–644.
30. K. Smid, G. Zoric, I.S. Pandzic, [Huge]: Universal architecture for statistically based human gesturing, in: Proceedings of the 6th International Conference on Intelligent Virtual Agents IVA 2006, 2006, pp. 256–269.
31. P. Ekman, About brows: emotional and conversational signals, in: M. von Cranach, K. Foppa, W. Lepenies, D. Ploog (Eds.), Human Ethology: Claims and Limits of a New Discipline, Cambridge University Press, Cambridge, 1979, pp. 169–202.
32. N. Chovil, Discourse-oriented facial displays in conversation, Res. Lang. Soc. Interact. 25 (1991) 163–194.
33. J. Cosnier, Les gestes de la question, in: Kerbrat-Orecchioni, (Ed.), La Question, Presses Universitaires de Lyon, 1991, pp. 163–171.
34. C. Cavé, I. Guaïtella, R. Bertrand, S. Santi, F. Harlay, R. Espesser, About the relationship between eyebrow movements and f0 variations, in: H.T. Bunnell, W. Idsardi (Eds.), ICSLP'96, 1996, pp. 2175–2178.
35. T. Kuratate, K. Munhall, P. Rubin, E. Vatikiotis-Bateson, H. Yehia, Audio-visual synthesis of talking faces from speech production correlates, in: EuroSpeech99, vol. 3, 1999, pp. 1279–1282.

36. K.G. Munhall, J.A. Jones, D.E. Callan, E. Kuratate, T. Bateson, Visual prosody and speech intelligibility: head movement improves auditory speech perception, Psychol. Sci. 15 (2) (2004) 133–137.

37. B. Granström, D. House, M. Lundeberg, Eyebrow movements as a cue to prominence, in: The Third Swedish Symposium on Multimodal Communication, 1999.

38. D. House, J. Beskow, B. Granström, Timing and interaction of visual cues for prominence in audiovisual speech perception, in: Proceedings of Eurospeech, 2001.

39. M. Argyle, M. Cook, Gaze and Mutual Gaze, Cambridge University Press, 1976.

40. I. Poggi, C. Pelachaud, Signals and meanings of gaze in animated faces, in: Language, Vision and Music: Selected Papers from the 8th International Workshop on the Cognitive Science of Natural Language Processing, John Benjamins, 2002, pp. 133–144.

41. C. Pelachaud, N.I. Badler, M. Steedman, Generating facial expressions for speech, Cogn. Sci. 20 (1996) 1–46.

42. M. Lundeberg, J. Beskow, Developing a 3D-agent for the august dialogue system, in: Proceedings of AVSP Workshop, 1999, pp. 151–154.

43. J. Lee, S. Marsella, Nonverbal behavior generator for embodied conversational agents, in: Intelligent Virtual Agents, 2006, pp. 243–255.

44. J. Cassell, Embodied Conversational Agents, The MIT Press, 2000.

45. D. Heylen, S. Kopp, S.C. Marsella, C. Pelachaud, H. Vilhjálmsson, The next step towards a function markup language, in: IVA'08: Proceedings of the 8th International Conference on Intelligent Virtual Agents, Springer, Berlin, Heidelberg, 2008, pp. 270–280.

46. P.C, B.M, Computational model of believable conversational agents, in: J.G. Carbonell, J. Siekmann (Eds.), Communications in Multiagent Systems, Springer Berlin, Heidelberg, 2003, pp. 300–317.

47. J. Cassell, C. Pelachaud, N. Badler, M. Steedman, B. Achorn, T. Becket, et al., Animated conversation: rule-based generation of facial expression, gesture & spoken intonation for multiple conversational agents, Comput. Graph. 28 (Annual Conference Series) (1994) 413–420.

48. P. Kenny, A. Hartholt, J. Gratch, W. Swartout, D. Traum, S. Marsella, et al., Building interactive virtual humans for training environment, in: Interservice/Industry Training, Simulation and Education Conference (I/ITSEC) 2007, 2007.

49. Microsoft speech API, microsoft speech technologies.

50. M. Schroder, J. Trouvain, The German text-to-speech synthesis system MARY: a tool for research, development and teaching, Int. J. Speech Technol. 6 (2003) 365–377.

51. P.A. Taylor, A. Black, R. Caley, The architecture of the festival speech synthesis system, in: The Third ESCA Workshop in Speech Synthesis, 1998, pp. 147–151.

52. K. van Deemter, B. Krenn, P. Piwek, M. Klesen, M. Schröder, S. Baumann, Fully generated scripted dialogue for embodied agents, Artif. Intell. 172 (10) (2008) 1219–1244.

53. B. Hartmann, M. Mancini, C. Pelachaud, Formational parameters and adaptive prototype instantiation for mpeg-4 compliant gesture synthesis, in: Proceedings of the Computer Animation, 2002, pp. 111–119.

54. A. Cerekovic, T. Pejsa, I.S. Pandzic, Real actor: character animation and multi-modal behavior realization system, in: Proceedings of Intelligent Virtual Agents IVA, 2009.

Chapter 14

Interactive Representations of Multimodal Databases

Stéphane Marchand-Maillet, Donn Morrison,
Enikö Szekely, and Eric Bruno
University of Geneva, Switzerland

14.1 INTRODUCTION

Several attempts to develop large-scale multimedia search engines have been described recently. Images and videos are mainly considered from their (audio)visual content viewpoints [1], and consensus

Multimodal Signal Processing, ISBN: 9780123748256

technologies have emerged for both feature extraction and indexing. Associated metadata comes generally in a textual form, for which feature extraction and indexing procedures are now mature and well established [2].

Most multimedia management frameworks are directed towards search operations. They use query-by-example (QBE) or concept-based query modes to retrieve relevant items within a collection. A number of advances have been made recently in integrating browsing and navigation, either as a complement to these search operations or as investigation tools in themselves. In this chapter, we review the construction of such interactive platforms and their potential exploitation for the long-term management of multimedia collections. In Section 14.2, we recall how multimedia items may be attached to viable and meaningful representations for further processing and how they may be associated with efficient access modes via indexing. In Section 14.3, we then review proposals for constructing browsing tools from these representations, with respect to the attached indexing strategies and their inherent limitations. Finally, in Section 14.4, we review how interaction gathered from such platforms may be exploited in a long-term scenario to gather semantic information and therefore strengthen their search and content abstraction capabilities.

14.2 MULTIMODAL DATA REPRESENTATION

Feature extraction is a necessary preprocessing step to map the data onto a form accessible by an automated process. It is also a unifying step, allowing to subsequently consider multimedia as a generic media (be it text, audio or visual) that holds common properties, possibly complemented to origin-specific features or properties. Features can be divided into low-level and high-level categories. The lowest level corresponds to any processing of the raw signals that do not imply any semantic decision, while high-level features involve semantic concepts like the presence of a face or concepts belonging to wider taxonomies, such as that proposed in LSCOM [3] or the Open Directory Project[1].

---

1. www.dmoz.org

Global visual attributes are generally related to colour, shape and texture content. For example, edge direction histogram, grid colour moment and Gabor texture are commonly used in content-based retrieval systems [4]. In addition to holistic descriptors, object-based attributes are also extensively used. Local descriptors, such as SIFT features [5], SURF [6] or Haar wavelet decomposition [7] are now well-established attributes for object and face detection/recognition [8].

Processing audio signals grants access to a rather complete picture of the audio content. Extracting both temporal and spectral attributes provides audio similarity [9] and higher-level description (e.g., speech transcripts [10], music genre [11, 12] and more general audio categories [13]) when combined to statistical models such as HMM [14]. The associated text (speech transcripts, meta-data and descriptions) then itself represents a low-level feature from which semantics (e.g., name entities) might be estimated [15].

Efficient access strategies must be installed on top of the extracted attributes. Indexes are the link between search interfaces and the data represented by its features. Indexing structures have to support severe constraints from these two sides: while flexible and fast access services are required by the retrieval components, features have to undergo minimal distortions when ingested into indexes in order to preserve document similarities as much as possible.

The high dimensionality of the resulting feature representation is an additional constraint limiting the spectrum of available indexing structures. Eligible similarity indexing structures for high-dimensional data include hashing [16], tree structures [17–20], space embedding based on fast algorithms [21, 22] and other space quantisation techniques like VA files [23] and their kernel extensions [24, 25]. Most of these structures provide approximate similarity searches. They have demonstrated their ability to index very large collections [26, 27], but leave the user with a very simple 'more like this' search functionality limited to the k-nearest neighbours.

The 'bag of visual words' model associated with inverted file structures has been introduced by the Computer Vision community [8]. It is now a very popular approach in content-based search [28–32]. Inspired from text retrieval, the idea relies on the construction of a codebook over the features. The codewords are estimated by some clustering techniques, where each cluster represents a visual term. Similarly to text documents, images are composed of a number of visual terms and may thus be indexed through a well-tried text indexing structure (term frequency-inverse document frequency (TF-IDF) weights stored into inverted files [33]). The inverted file provides a fast access to the data when the query is limited to a few words manually provided by users. Unfortunately, this strategy is less effective for image retrieval where words are visual patterns and queries made out of several example images composed of hundreds or thousands of words [34]. A second limitation comes from the construction of the visual codebook. The arbitrary number of visual terms is a sensitive parameter as it determines the trade-off between an over-fitting and a non-discriminative content description. It has been shown recently that this number is mostly context dependent [31], and therefore suboptimal when addressing the problem of broad and generic visual content.

An alternative approach, related to the current study, is to represent documents according to their similarities (related to one or several features) to the other documents rather than to a feature vector. Considering a collection of documents, the similarity-based representation, stored in (dis)similarity matrices or some distance-based indexing structures [22, 35, 36], characterises the content of an element of the collection relatively to a part of or the whole collection. Recent studies have been published for document retrieval and collection browsing using pre-computed similarities. In [37], Boldareva et al., proposed to index elements relatively to their closest neighbours, i.e., those who have the best probabilities to belong to the same class. This provides them with a sparse association graph structuring the multimedia collection and allowing fast retrieval of data. In [38], the idea of nearest neighbour networks is extended by creating edges for every combination of features. The resulting graph, called NN^k, permits browsing the data collection from various viewpoints corresponding to the multiple features (see Section 14.3.1). In [39, 40], dissimilarity spaces are constructed based on item interdistances and exploited to homogenise features and perform optimal information fusion at search time. The size of current multimedia collections makes the management of the indexing structures complex. Distributed and parallel processing is certainly a key-issue to really attain the large-scale objective. Its role is to divide the computational effort of indexing and retrieval operations over a number of CPUs and memory devices. Although feature extraction is somewhat easily distributed over several CPUs with a coarse-grain strategy, obtaining efficiently distributed indexing and learning procedures is more challenging, especially in the context of multimodal retrieval.

The large-scale distributed indexing research field has been investigated for more than a decade. As a result, many distributed and parallel structures allowing nearest-neighbour (NN) search in sublinear complexity have been proposed, and they are routinely used nowadays in commercial applications [41]. Distributed inverted files [42, 43], parallel VA files [44] and parallel metric trees [45] are highly relevant approaches here.

14.3   MULTIMODAL DATA ACCESS

On the basis of aforementioned multimodal representations, search and retrieval operations may be initiated. Many current information management systems are centred on the notion of a query. This is true over the Web (with all classical Web Search Engines), and for Digital Libraries. In the domain of multimedia, available commercial applications propose rather simple management services whereas research prototypes are also looking at responding to queries. The notion of browsing may be used to extend query-based operations or may simply form an alternative access mode in several possible contexts that we detail in the following sections.

14.3.1   Browsing as Extension of the Query Formulation Mechanism

In the most general case, multimedia browsing is designed to supplement search operations. This comes from the fact that the multimedia querying systems largely demonstrate their capabilities using the QBE scenario, which hardly corresponds to a usable scenario. Multimedia search systems are mostly based on content similarity. Hence, to fulfil an information need, the user must express it with respect to relevant (positive) and non-relevant (negative) examples [46]. From there, some form of learning is performed, to retrieve the documents that are the most similar to the combination of relevant examples and dissimilar to the combination of non-relevant examples (see Section 14.4.1). The question then arises of how to find the initial examples themselves. Researchers have therefore investigated new tools and protocols for the discovery of relevant bootstrapping examples. These tools often take the form of browsing interfaces whose aim is to help the user exploring the information space to locate the sought items.

The initial query step of most QBE-based systems consists in showing images in random sequential order over a 2D grid [46]. This follows the idea that a random sampling will be representative of the collection content and allow for choosing relevant examples. However, the chance for gathering sufficient relevant examples is low and much effort must be spent in guiding the system towards the relevant region of information space where the sought items may lie. Similarity-based visualisation [47–55] organises images with respect to their perceived similarities. Similarity is mapped onto the notion of distance so that a dimension reduction technique may generate a 2D or 3D space representation where images may be organised. It is further known that high dimensionality has an impact on the meaningfulness of the distances defined [56]. This is known as the curse of dimensionality (see Chapter 2) and several results can be proven that there is a need for avoiding high-dimensional spaces, where possible. A number of methods exist to achieve dimension reduction. We do not detail the list and principles here but refer the reader to [57–60] for thorough reviews on the topic.

Figure 14.1a illustrates the organisation of an image collection based on colour information using the HDME dimension reduction [58]. This type of display may be used to capture feedback by letting the user reorganise or validate the displayed images. Figure 14.1b shows a screenshot of the interface of the El Niño system [61] with such a configuration.

Specific devices may then be used to perform search operations. Figure 14.2 shows operators sitting around an interactive table for

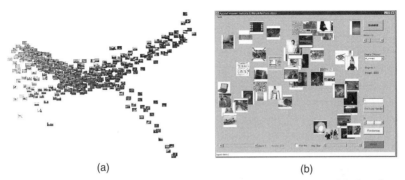

(a)   (b)

FIGURE 14.1   (a) Dimension reduction over a database of images and (b) interface of the El Niño system [61].

(a)   (b)

FIGURE 14.2   The PDH table and its artistic rendering (from [50]).

handling personal photo collections [50]. Figure 14.3 shows an operator manipulating images in front of a large multi-touch display[2]. Alternative item organisations are also proposed such as the Ostensive Browsers (see Figure 14.4 and [62]) and interfaces associated to the NN^k paradigm [38].

All these interfaces have in common the fact of placing multimedia retrieval much closer to human factors and therefore require specific evaluation procedures, as detailed in Section 14.3.4. Although somewhat different, it is worth mentioning here the development of

---

2. From http://www.perceptivepixel.com

FIGURE 14.3 Manipulating images over touch-enabled devices.

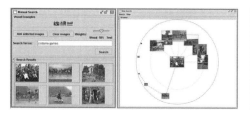

FIGURE 14.4 The Ostensive Browsers [100].

the Target Search browsers. Using QBE-based search a user may formulate a query of the type 'show me everything that is similar to this (and not similar to that)' and thus characterise a set of images, using Target Search, the user is looking for a specific image (s)he knows is in the collection. By iteratively providing relative feedback on whether some of the current images are closer to the target than others, the user is guided to the target image. This departs from the QBE-based search where the feedback is absolute ('this image is similar to what I look for, whatever the context'). In that sense, Bayesian search tools may be considered as focused collection browsers. In this category, the PicHunter Bayesian browser [63] is one of the initial developments. It has been enhanced with refocusing capabilities in [64] via the development of the Tracker system.

14.3.2   Browsing for the Exploration of the Content Space

In the aforementioned works, browsing is seen as an alternative to the random picking of initial examples for the QBE paradigm. Here, we look at browsing from a different point of view. In this set-up, the user aims at overviewing the collection with no specific information need. Simply, (s)he wishes to acquire a representative view on the collection. In some respect, the aforementioned developments may be included into this category as overviews of the sub-collection representative to the query in question. In [65], specific presentation layouts are proposed and evaluated (see also section 14.3.4). The interface aims at enhancing the classical grid layout by organising related image groups around a central group (see Figure 14.5a). Somewhat similar is the earlier development of PhotoMesa [66], which aims at browsing image hierarchies using treemaps (Figure 14.5b). Hierarchies are also studied in depth in the Muvis system, both for indexing and browsing via the Hierarchical Cluster Tree (HCT) structure [67]. In Figure 14.6, an example of hierarchical browsing of a relatively small image collection (1000 images) is displayed.

In [68], the alternative idea of linearising the image collection is presented. The collection is spanned by two space-filling curves that allow for aligning the images along two intersecting 1D paths. The reason for allowing two paths is that while two neighbouring

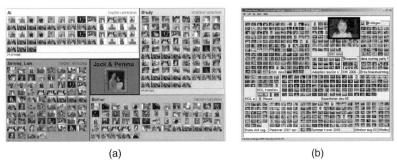

(a)   (b)

FIGURE 14.5   (a) Bi-level radial layout [65]. (b) Screenshot of PhotoMesa, based on TreeMaps [66].

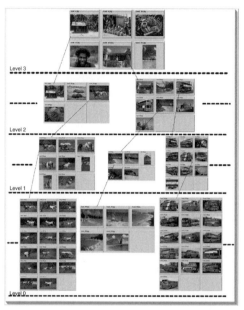

FIGURE 14.6   An example of HCT-based hierarchical navigation [67] on the 1k Corel image collection.

points on a space-filling curve are neighbours in the original space, the converse is not guaranteed to be true. Hence, two neighbouring points in the original space may end up far apart on the path. The use of two interweaved curves may alleviate this shortcoming. At every image, each of the two paths may be followed in either of the two directions so that at every image, four directions are allowed. A browser shown in Figure 14.7 is proposed to materialise this visit.

In [69], the principle of Collection Guiding is introduced. Given the collection of images, a path is created so as to 'guide' the visit of the collection. For that purpose, image inter-similarity is computed and the path is created via a Travelling Salesman tour of the collection. The aim is to provide the user with an exploration strategy based on a minimal variation of content at every step. This implicitly provides a dimension reduction method from a high-dimensional feature space to a linear ordering. In turn, this allows for emulating sort operations on the collection, as illustrated in Figure 14.8.

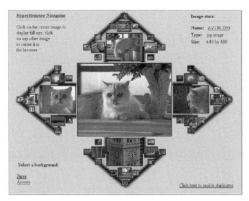

FIGURE 14.7   Multi-linearisation browser [68].

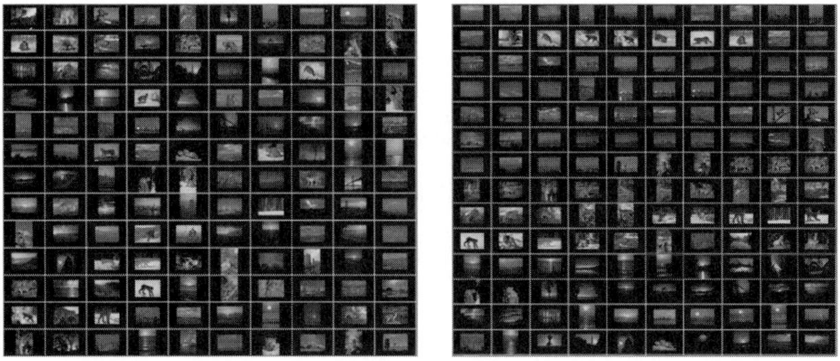

(a)   (b)

FIGURE 14.8   Image sorting via the Collection Guide (a) random order (b) sorted list.

The Collection Guide provides also several multi-dimensional arrangements (see Figure 14.9). However, it is clear that these (as the ones presented in the earlier section) are conditioned to the quality of the dimension reduction strategy. In [57], the underlying data cluster structure is accounted for so as to deploy valid dimension reduction operations (see also [57–60]). Combining many of the above principles derived from the analysis of this state of the art, an information browser is proposed in [70] as complement to a main search interface. This work follows the idea that searching and relevance feedback help in exploring local portions of the information space, whereas

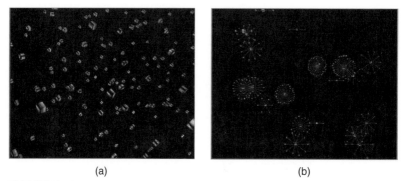

(a)   (b)

FIGURE 14.9 Examples of displays provided by the Collection Guide. (a) Planet metaphor; (b) generic 3D mapping.

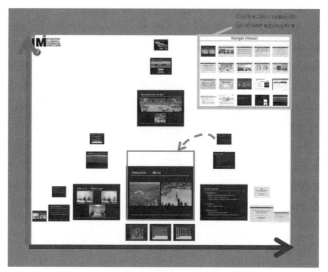

FIGURE 14.10 The (IM)2 collection browser.

browsing should help the user getting both a global overview of the information space at hand and provide the user with a clear and efficient navigation strategy. The idea of the Collection Guide is thus mixed with that of linearisation and faceted browsing to obtain an information browser starting from a specific document and linking, out of several possible dimensions [57], to other documents close to that dimension. Clicking on any of the non-central documents would bring it to the central place with its associated context and thus trigger a move within the information space (see Figure 14.10).

Finally, at the border between exploration and search, opportunistic search is 'characterised by uncertainty in user's initial information needs and subsequent modification of search queries to improve on the results' [71, 72]. In [71], the authors present a visual interface using semantic fisheye views to allow the interaction over a collection of annotated images. Figure 14.11 displays interfaces associated with this concept.

Faceted browsing [73], oriented towards search is also at the limit between exploration and querying as it also for filtering a collection while smoothly and interactively constructing complex queries. Figure 14.12 displays an example application of Faceted Search using the Flamenco toolbox for a collection of annotated images.

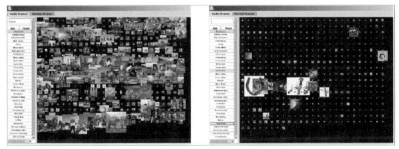

FIGURE 14.11   Displays associated with the opportunistic search mechanism (from [71] and [72]).

FIGURE 14.12   UC Berkeley Architecture Image Library (Flamenco toolbox).

14.3.3 Alternative Representations

Although retrieval and browsing are in general passive to the collection (i.e., the collection stays as it is), these operations may also be used to enrich the collection content. In [74], authors have reviewed and proposed several models that allow for the semantic augmentation of multimedia collections via interacting with them. This follows the line of the Semantic Web and associated domains of knowledge management. In this line, the work proposed in [75] relates ontology management and image description. In Section 14.4, we further present strategies for the semantic enrichment of media collecions.

Associated with the concepts of exploration and browsing is the concept of summarisation. Summarisation is an approach commonly taken for presenting large content and involves a clear understanding of the collection diversity for performing sampling. The most common way of performing sampling is to use the underlying statistics of the collection. Typically, within the feature space, local density is analysed. Dense regions of this space will be represented by several items whereas sparse regions will mostly be ignored within the representation. More formally, strategies such as Vector Quantisation (VQ) may be used to split the space into cells and only consider cell representatives. k-means clustering is one of the most popular VQ techniques. A geometrical interpretation of VQ is that of defining a Voronoi (Dirichlet) tessellation of the feature space such that each cell contains a cluster of data points and each centroid is the seed of the corresponding cell. This tessellation is optimal in terms of minimising some given cost function, embedding the assumption over the properties of the similarity measurement function in the image representation space.

14.3.4 Evaluation

In [76], it is analysed how browsing and the more general fact of providing an efficient interface to information systems is often listed as one of 'Top-ten' problems in several fields (e.g., information retrieval [77], visualisation and virtual reality). A new top-ten list of problems in the domain is created including benchmarking and evaluation.

First, the majority of browsing tools proposed in the literature organises their content using low-level features such as colour or texture. Rorissa et al. [78] demonstrates via several user studies that this is relevant and that features may indeed be used as a basis for visualisation and hence browsing. There are numerous efforts to benchmark information retrieval as a problem with a well-posed formulation. When including the quality of the interface or performance of the interaction with the information system, things are however less clear. The fact of embarking human factors in the context make the formulation less definite and prevents the automation of the performance measures (see e.g., [79]). Several attempts to propose evaluation protocols and frameworks have nevertheless taken place [80–82]. Some particular aspects such as zooming [83] and presentation [65, 80] have been the focus of attention for some works. While systematic retrieval performance evaluation is possible using ground truth and measures such as Precision and Recall, having reliable performance evaluation of interfaces and interactive tools requires long-term efforts and heavy protocols. It is certainly an area where developments should be made to formally validate findings. It is often a strong asset of private companies which carefully invest in a user-based testing to validate tools that are simpler but more robust than most research prototypes.

14.3.5   Commercial Impact

Image browsing is of high-commercial interest because it provides an added value over a collection of data. We list here some known commercial browsers:

- Microsoft's picture manager (filmstrip mode) is the simplest representation that can be created. It exploits a linear organisation of the data. In the context of its usage, linearisation is made on simple metadata, which lends itself to the ordering (e.g., temporal or alphabetical order).
- Google's Picasa (timeline mode) also exploits the linear timeline to arrange a photo collection. An interesting feature is the near-1D organisation whereby groups of pictures are arranged along the path (as opposed to aligning single pictures).

- Flickr's geotagged image browser exploits the planar nature of geographic data to arrange pictures.

As a complement, the following are some U.S. patents related to image browsing:

- 6233367 Multilinearisation data structure for image browsing.
- 6636847 Exhaustive search system and method using space-filling curves.
- 6907141 Image data sorting device and image data sorting method.
- 7003518 Multimedia searching method using histogram.
- 7016553 Linearised data structure ordering images based on their attributes.
- 7131059 Scalably presenting a collection of media objects.
- 7149755 Presenting a collection of media objects.

14.4 GAINING SEMANTIC FROM USER INTERACTION

Search and browsing operations are by nature interactive. Behind this interaction lies the intention of the user (his/her 'mental image'). Hence, be it directly or indirectly, semantic information may be extracted from that interaction. In Section 14.4.1, we review how interactive search mechanisms may be translated into online learning procedures. In Section 14.4.2, we take the perspective of indirectly exploiting group interaction (user sessions over a collection) to enrich the collection with further semantic interaction.

14.4.1 Multimodal Interactive Retrieval

Determining semantic concepts by allowing users to iteratively and interactively refine their queries is a key issue in multimedia content-based retrieval. For instance, the Relevance Feedback loop allows to build complex queries made out of documents marked as positive and negative examples [84]. From this training set, a learning process has to create a model of the sought concept from a set of data features to finally provide relevant documents to the user. The success of this search strategy relies mainly on the representation spaces within which the data is embedded as well as on the learn-ing algorithms operating in those spaces. These two issues are also intrinsically related to the problem of adequately fusing informa-tion arising from different sources. Various aspects of these problems have been studied with success for the last few years. These include works on machine-learning strategies such as active learning [85], imbalance classification algorithms [86], automatic kernel setting [87] or automatic labelling of training data [88]. Theoretical and experimental investigations have been achieved to determine opti-mal strategies for multimodal fusion: Kittler et al. [89] and Duin [90] studied different rules for classifier combination; Wu et al. [91] pro-pose the super-kernel fusion to determine optimal combination of features for video retrieval. In [92], Maximum Entropy, Boosting and SVM algorithms are compared to fuse audio-visual features. Multi-graph learning approaches [93] and latent semantic fusion [94] have been recently proposed for image and video retrieval and annotation. A number of further relevant references may be found in the Lecture Notes series on Multiple Classifier Systems [95].

While providing adaptive and multimodal retrieval functional-ities, such sophisticated algorithms are limited in their ability to handle large collections. Re-ranking algorithms have been recently proposed to address this problem. The strategy consists of involv-ing sophisticated learning-based fusion algorithms on a small set of documents previously shortlisted with simpler and less-effective approaches [32, 96]. Another approach consists in limiting complex retrieval computation to the most popular queries asked by users [16] (it is worth noting that this strategy is feasible when popular queries are available).

14.4.2   Crowdsourcing

Increasingly, research is moving towards exploiting crowdsourcing to better aid filtering and retrieval. Crowdsourcing is a moniker, which describes the outsourcing of a problem or task to a large number of users in an attempt at finding a solution [97]. Although the def-inition lends itself to an explicit arrangement of the distribution of problem workload, such as the Netflix Prize [98], it also covers implic-itly sourced user-power, where tasks are less defined and subtle, an example being the reCAPTCHA effort to fight spam while help-ing optical character recognition in book scanning [99]. The latter system works by presenting examples of words that could not be cor-rectly read by optical character recognition (OCR) as tests, to discern between real humans and spam bots (spam-generating scripts).

Many variations have appeared in the literature since Relevance Feedback has been declared useful but it was only until recently that research began focusing on using long-term learning on cumulative relevance feedback instances [100–102]. Although the historical data provided by relevance feedback is used extensively in the following literature, the actual relevance feedback algorithms and techniques in the short-term sense are not reviewed here. For a thorough review on relevance feedback for short-term learning, the reader is directed to both [103] and Chapter 11 of [2] which specifically covers user interfaces and interaction.

Because of the complex nature of the problem, many areas have been explored to bridge the semantic gap using long-term learning. The core idea is to extend the traditional relevance feedback model such that it persists over future queries. The motivation behind this idea can be seen in the probability distribution that emerges from user interaction. In many natural settings, including human interaction, a power law distribution is observed (see Figure 14.13). In information

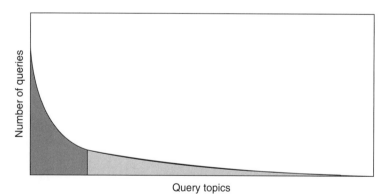

FIGURE 14.13 An illustrative example of the long tail of user queries in an infor-mation retrieval system. The majority of the searches comprise only a small portion of the distribution, leaving a tail composed of infrequent searches.

retrieval, it has been shown that this distribution, also known as the 80–20 rule, yields queries in which the most frequent fall in the first 20% and the less frequent are distributed along the *long tail* – the last 80% [104]. If 80% of users search for 20% of the information, there is a large amount of duplication involved. By learning past queries and using them in the future, a great deal of work (both that of the system and the user) can be saved.

One of the first studies that looked at using inter-query learning to aid future queries is [100]. The authors analysed the logs of queries using a demonstration system over a long period of time and used this information to update TF-IDF feature weightings in the low-level feature index.

In [101], a general framework is described which annotates the images in a collection using relevance feedback instances. As a user browses an image database using a retrieval system, providing relevance feedback as the query progresses, the system automatically annotates images using the relationships described by the user. In [105], the authors combine inter-query learning with traditional low-level image features to build semantic similarities between images for use in later retrieval sessions. The similarity model between the request and target images are refined during a standard relevance feedback process for the current session. Similarly, in [106], a statistical correlation model is built to create semantic relationships between images based on the co-occurrence frequency that images are rated relevant to a query. These relationships are also fused with low-level features to propagate the annotations onto unseen images.

Inter-query learning is used in [102] to improve the accuracy of a retrieval system with latent semantic analysis (LSI). Random queries were created, and two sessions of relevance feedback were conducted to generate the long-term data to be processed by LSI. From experiments on different levels of data, they conclude that LSI is robust to a lack of data quality but is highly dependent on the sparsity of interaction data. In another study, authors use long-term learning in the PicSOM retrieval system [107]. PicSOM is based on multiple parallel tree-structured *self-organising maps* (SOMs) and uses MPEG7 content descriptors for features. The authors claim that by the use of SOMs the system automatically picks the most relevant features.

Relevance feedback is used in [108] to generate a semantic space on which a support vector machine is trained. Low-level features are used in conjunction with the long-term relevance feedback data to improve performance in the MiAlbum image retrieval system. Artificial relevance feedback data was generated by running simulated queries on a database of categorised images. The positive and negative examples were taken from the top three correct and top three incorrect results, respectively.

In [109], long-term user interaction with a relevance feedback system is used to make better semantic judgements on unlabelled images for the purpose of image annotation. Relationships between images that are created during relevance feedback can denote similar or dissimilar concepts. The authors also try to improve the learning of semantic features by 'a moving of the feature vectors' around a group of concept points, without specifically computing the concept points. The idea is to cluster the vectors around the concept centres.

Finally, Markov random walks are employed in [110] on a large bipartite click graph of queries and documents (images) collected from popular online search engines. By following walks either backward or forward from the query on the graph, document clusters can be found for associated search keywords.

14.5 CONCLUSION AND DISCUSSION

Multimedia browsing comes as a complement to query-based search. This is valuable, due to the imperfect nature of content understanding and representation, due principally to the so-called semantic gap. Browsing is also interesting to resolve the problem of the user's uncertainty in formulating an information need. Opportunistic search and faceted browsing are example of principles and applications that bridge search and navigation.

The aforementioned analysis shows that, as a complement to classical retrieval systems, browsing and navigation should be differentiated. It is suggested here that browsing is directed towards an objective (information need) and thus indirectly relates to searching and acts at the document scale. As such, browsing is seen as assistance within similarity-based search systems, where the QBE paradigm is often deficient. Browsing should be differentiated from navigating where the aim is the understanding of the collection content. Navigation-based systems thus use an absolute (global) modelling of the collection and include a global notion of similarity (i.e., driven by generic feature). This is to be opposed to browsing systems, which use a notion of similarity based on the context of the neighbourhood of the sought items (i.e., the interpretation of the collection is made at the light of the sought items).

Collection browsing imposes to focus on user interaction and thus the interface design and evaluation. This refers to the work done by the Human Factors (HCI) community, which is somewhat regrettably not enough inter-weaved with the Information Retrieval and management community.

Finally, while most of the examples taken here refer to image management (seen as multimodal data due to the presence of text that may be exploited in display or retrieval), browsing and navigation, as an extension and complement to searching, can also be applied to other media such as audio (music, e.g., [111]) and video (e.g., [112, 113]). These temporal media offer a temporal dimension that directly lends itself to exploration and thus makes browsing an obvious tool to use.

Acknowledgments

The support of SNF via subsidy 200020-121842 in parallel with the NCCR(IM)2 is acknowledged here.

REFERENCES

1. R. Datta, D. Joshi, J. Li, J.Z. Wang, Image retrieval: ideas, influences, and trends of the new age, ACM Comput. Surv. 40 (2) (2008) 1–60.
2. R. Baeza-Yates, N.-R. Berthier, Modern Information Retrieval, Addison-Wesley, Essex, England, 1999.
3. M. Naphade, J.R. Smith, J. Tesic, S.-F. Chang, W. Hsu, L. Kennedy, et al., Large-scale concept ontology for multimedia, IEEE MultiMedia, 13 (3) (2006) 86–91.
4. S.-F. Chang, D. Ellis, W. Jiang, K. Lee, A. Yanagawa, A.C. Loui, et al., Large-scale multimodal semantic concept detection for consumer video, in: MIR '07: Proceedings of the International Workshop on Multimedia Information Retrieval, ACM Press, New York, NY, USA, 2007, pp. 255–264.

5. D. Lowe, Object recognition from local scale invariant features, in: Proceedings of the International Conference in Computer Vision, ICCV'99, Corfu, 1999, pp. 1150–1157.

6. H. Bay, T. Tuytelaars, L. Van Gool, Surf: speeded up robust features, in: 9th European Conference on Computer Vision, pp. 14–28, Graz Austria, May 2006.

7. P. Viola, M. Jones, Robust real-time face detection, Int. J. Comput. Vision (IJCV) 57 (2) (2004) 137–154.

8. J. Ponce, M. Hebert, C. Schmid, A. Zisserman (Eds.), Toward Category Level Object Recognition, vol. 4170 of Lecture Notes in Computer Science, Springer, 2006.

9. J. Gu, L. Lu, H. Zhang, J. Yang, Dominant feature vectors based audio similarity measure, in: PCM, number 2, 2004, pp. 890–897.

10. J. Gauvain, L. Lamel, G. Adda, The limsi broadcast news transcription system, Speech Commun. 37 (1–2) (2002) 89–108.

11. J. Arenas-Garca, A. Meng, K.B. Petersen, T.L. Schiøler, L.K. Hansen, J. Larsen, Unveiling music structure via PLSA similarity fusion, in: IEEE International Workshop on IEEE International Workshop on Machine Learning for Signal Processing, IEEE Press, August 2007, pp. 419–424.

12. H. Homburg, I. Mierswa, K. Morik, B. Möller, M. Wurst, A benchmark dataset for audio classification and clustering, in: Proceedings of ISMIR, 2005, pp. 528–531.

13. M.F. Mckinney, Features for audio and music classification, in: Proceedings of the International Symposium on Music Information Retrieval, 2003, pp. 151–158.

14. L. Rabiner, A tutorial on HMM and selected applications in speech recognition, Proc. IEEE 77 (2) (1989) 257–286.

15. E. Alfonseca, S. Man, S. Manandhar, an unsupervised method for general named entity recognition and automated concept discovery, in: Proceedings of the 1st International Conference on General WordNet, Mysore, India, p. 303, 2002.

16. Y. Jing, S. Baluja, Visualrank: applying pagerank to large-scale image search, IEEE Trans. Pattern Anal. Mach. Intell. 30 (11) (2008) 1877–1890.

17. A. Guttman, R-trees: a dynamic index structure for spatial searching, in: SIGMOD '84: Proceedings of the 1984 ACM SIGMOD International Conference on Management of Data, ACM Press, New York, NY, USA, 1984, pp. 47–57.

18. P. Zezula, P. Savino, G. Amato, F. Rabitti, Approximate similarity retrieval with m-trees, VLDB J. 7 (4) (1998) 275–293.

19. M.A. Nascimento, E. Tousidou, V. Chitkara, Y. Manolopoulos, Image indexing and retrieval using signature trees, Data Knowl. Eng. 43 (1) (2002) 57–77.

20. D. Novak, M. Batko, P. Zezula, Web-scale system for image similarity search: when the dreams are coming true, in: CBMI' 08, June 2008, pp. 446–453.

21. V. Athitsos, J. Alon, S. Sclaroff, G. Kollios, Boostmap: an embedding method for efficient nearest neighbor retrieval, IEEE Trans. Pattern Anal. Mach. Intell. 30(1) (2008) 89–104.

22. C. Faloutsos, K. Lin, FastMap: a fast algorithm for indexing, data-mining and visualization of traditional and multimedia datasets, in: Proceedings of the 1995 ACM SIGMOD International Conference on Management of Data, San Jose, California, May 22–25, 1995, pp. 163–174.

23. R. Weber, K. Böhm, Trading quality for time with nearest neighbor search, in: EDBT, 2000, pp. 21–35.

24. D.R. Heisterkamp, J. Peng, Kernel VA-files for relevance feedback retrieval, in: MMDB '03: Proceedings of the 1st ACM International Workshop on Multimedia Databases, ACM Press, New York, NY, USA, 2003, pp. 48–54.

25. D.R. Heisterkamp, J. Peng, Kernel vector approximation files for relevance feedback retrieval in large image databases, Multimedia Tools Appl. 26 (2) (2005) 175–189.

26. G. Amato, P. Savino, Approximate similarity search in metric spaces using inverted files, in: InfoScale '08: Proceedings of the 3rd International Conference on Scalable Information Systems, ICST, Brussels, Belgium, Belgium, 2008, pp. 1–10. ICST (Institute for Computer Sciences, Social-Informatics and Telecommunications Engineering).

27. J. Lai, Y. Liaw, J. Liu, Fast k-nearest-neighbor search based on projection and triangular inequality, Pattern Recognit. 40 (2) (2007) 351–359.

28. R. Fergus, P. Perona, A. Zisserman, A visual category filter for google images, in: ECCV (1), 2004, pp. 242–256.

29. E. Horster, R. Lienhart, Fusing local image descriptors for large-scale image retrieval, IEEE Comput. Soc. Conf. Comput. Vision Pattern Recognit. (2007) 1–8.

30. T. Quack, B. Leibe, L. Van Gool, World-scale mining of objects and events from community photo collections, in: CIVR '08: Proceedings of the 2008 International Conference on Content-Based Image and Video Retrieval, ACM Press, New York, NY, USA, 2008, pp. 47–56.

31. S.-F. Chang, J. He, Y.-G. Jiang, E. El Khoury, C.-W. Ngo, A. Yanagawa, Columbia University/VIREO-CityU/IRIT TRECVID2008 high-level feature extraction and interactive video search, in: NIST TRECVID Workshop, Gaithersburg, MD, November 2008. http://www.ce.columbia.edu/ln/dvmm/publicationPage//Author//Shih-Fu. Chang.html. Accessed 1 September 2009.

32. H. Jegou, M. Douze, C. Schmid, Hamming embedding and weak geometric consistency for large scale image search, in: A.Z. David Forsyth, P. Torr (Eds.), European Conference on Computer Vision, vol. 1 of LNCS, Springer, October 2008, pp. 304–317.

33. D.M. Squire, W. Müller, H. Müller, T. Pun, Content-based query of image databases: inspirations from text retrieval, Pattern Recognit. Lett. 21 (13–15) (2000) 1193–1198.

34. Z. Li, X. Xie, L. Zhang, W.-Y. Ma, Searching one billion web images by content: challenges and opportunities, in: MCAM, 2007, pp. 33–36.

35. E. Chávez, G. Navarro, R. Baeza-Yates, J. Marroquin, Searching in metric spaces, ACM Comput. Surv. 33 (3) (2001) 273–321.

36. J. Vleugels, R.C. Veltkamp, Efficient image retrieval through vantage objects, Pattern Recognit. 35 (1) (2002) 69–80.

37. L. Boldareva, D. Hiemstra, Interactive content-based retrieval using pre-computed object–object similarities, in: Conference on Image and Video Retrieval, CIVR'04, Dublin, Ireland, 2004, pp. 308–316.

38. D. Heesch, S. Rueger, NN^k networks for content-based image retrieval, in: 26th European Conference on Information Retrieval, Sunderland, UK, 2004, pp. 253–266.

39. E. Bruno, N. Moënne-Loccoz, S. Marchand-Maillet, Design of multimodal dis-similarity spaces for retrieval of multimedia documents, IEEE Trans. Pattern Anal. Mach. Intell. 30 (9) (2008) 1520–1533.

40. E. Bruno, S. Marchand-Maillet, Multimodal preference aggregation for multimedia information retrieval, J. Multimedia (JMM), 4 (5) (2009) 321–329.

41. S. Ghemawat, H. Gobioff, S.-T. Leung, The google file system, in: SOSP '03: Proceedings of the Nineteenth ACM Symposium on Operating Systems Principles, ACM Press, New York, NY, USA, 2003, pp. 29–43.

42. C. Stanfill, Partitioned posting files: a parallel inverted file structure for information retrieval, in: SIGIR '90: Proceedings of the 13th Annual International ACM SIGIR Conference on Research and Development in Information Retrieval, ACM Press, New York, NY, USA, 1990, pp. 413–428.

43. D.K. Lee, L. Ren, Document ranking on weight-partitioned signature files, ACM Trans. Inf. Syst. 14 (2) (1996) 109–137.

44. R. Weber, K. Boehm, H.-J. Schek, Interactive-time similarity search for large image collections using parallel va-files, Data Eng. Int. Conf., IEEE Computer Society (2000) p. 197.

45. A. Alpkocak, T. Danisman, T. Ulker, A parallel similarity search in high dimensional metric space using m-tree, in: IWCC '01: Proceedings of the NATO Advanced Research Workshop on Advanced Environments, Tools,

and Applications for Cluster Computing-Revised Papers, Springer-Verlag, London, UK, 2002, pp. 166–171.

46. A.W.M. Smeulders, M. Worring, S. Santini, A. Gupta, R. Jain, Content based image retrieval at the end of the early years, IEEE Trans. Pattern Anal. Mach. Intell. 22 (12) (2000) 1349–1380.

47. C. Chen, G. Gagaudakis, P. Rosin, Content-based image visualization, in: Proceedings of the International Conference on Information Visualization, 2000, pp. 13–18.

48. L. Cinque, S. Levialdi, A. Malizia, K. Olsen, A multidimensional image browser, J. Visual Language Comput. 9 (1) (1998) 103–117.

49. W. Leeuw, R. Liere, Visualization of multidimensional data using structure preserving projection methods, in: F. Post, G. Nielson, G. Bonneau (Eds.), Data Visualization: The State of the Art, Kluwer, 2003, pp. 213–224.

50. B. Moghaddam, Q. Tian, N. Lesh, C. Shen, T.S. Huang, Visualization and user-modeling for browsing personal photo libraries, Int. J. Comput. Vision 56 (1–2) (2004).

51. M. Nakazato, L. Manola, T.S. Huang, Imagegrouper: a group-oriented user interface for content-based image retrieval and digital image arrangement, J. Visual Lang. Comput. 14 (2003) 363–386.

52. G. Nguyen, M. Worring, Interactive access to large image collections using similarity-based visualization, J. Visual Lang. Comput. 19 (2) 203–204, 2006.

53. G. Nguyen, M. Worring, Optimization of interactive visual similarity based search, ACM TOMCCAP 4 (1) (2008) 1–23.

54. Y. Rubner, Perceptual Metrics for Image Database Navigation, PhD thesis, Stanford University, 1999.

55. C. Vertan, M. Ciuc, C. Fernandez–Maloigne, V. Buzuloiu, Browsing image databases with 2D image similarity scatter plots: update of the IRIS system, in: International Symposium on Communications, 2002, pp. 397–402.

56. C.C. Aggarwal, A. Hinneburg, D.A. Keim, On the surprising behavior of distance metrics in high dimensional space, Lecture Notes in Computer Science, 1973, pp. 420–434, 2001.

57. E. Székely, E. Bruno, S. Marchand-Maillet, Clustered multidimensional scaling for exploration in information retrieval, in: 1st International Conference on the Theory of Information Retrieval, pp. 95–104, Budapest, Hungary, 2007.

58. S. Marchand-Maillet, E. Szekely, E. Bruno, Collection guiding: review of main strategies for multimedia collection browsing, Technical Report 08.01, Viper group – University of Geneva, 2008.

59. I. Borg, P. Groenen, Modern Multidimensional Scaling: Theory and Application, Springer, 2005.

60. M.A. Carreira-Perpinan, A review of dimension reduction techniques, Technical Report CS-96-09, Dept. of Computer Science, University of Sheffield, UK, 1996.

61. S. Santini, A. Gupta, R. Jain, Emergent semantics through interaction in image databases, IEEE Trans. Knowl. Data Eng. 13 (3) (2001) 337–351.

62. J. Urban, J. Jose, An adaptive technique for content-based image retrieval, Multimedia Tools Appl. 31 (1) (2006) 1–28.

63. I. Cox, M. Miller, T. Minka, T. Papathornas, P. Yianilos, PicHunter: theory, implementation, and psychophysical experiments, IEEE Trans. Image Process. 9 (1) (2000) 20–37.

64. W. Müller, Hunting moving targets: an extension to Bayesian methods in multimedia databases, in: Multimedia Storage and Archiving Systems IV (VV02), vol. 3846 of SPIE Proceedings, pp. 328–337, 1999.

65. J. Kustanowitz, B. Shneiderman, Meaningful presentations of photo libraries: rationale and applications of bi-level radial quantum layouts, in: ACM/IEEE Joint Conference on Digital Libraries, 2005, pp. 188–196.

66. D. Bederson, Photomesa: a zoomable image browser using quantum treemaps and bubblemaps, in: ACM Symposium on User Interface Software and Technology, CHI Letters, vol. 3, 2001, pp. 71–80.

67. M. Kiranyaz, M. Gabbouj, Hierarchical cellular tree: an efficient indexing scheme for content-based retrieval on multimedia databases, IEEE Trans. Multimedia 9 (1) (2007) 102–119.

68. S. Craver, B.-L. Yeo, M. Yeung, Multi-linearisation data structure for image browsing, in: SPIE Conference on Storage and Retrieval for Image and Video Databases VII, January 1999, pp. 155–166.

69. S. Marchand-Maillet, E. Bruno, Collection guiding: a new framework for handling large multimedia collections, in: Seventh International Workshop on Audio-Visual Content and Information Visualization in Digital Libraries (AVIVDiLib'05), pp. 180–183, Cortona, Italy, 2005.

70. S. Marchand-Maillet, E. Szekely, E. Bruno, Optimizing strategies for the exploration of social-networks and associated data collections, in: Proceedings of the International Workshop on Image Analysis for Multimedia Interactive Services (WIAMIS'09) – Special Session on "People, Pixels, Peers: Interactive Content in Social Networks", pp. 29–32, London, UK, 2009. (Invited).

71. P. Pu, P. Janecek, Visual interfaces for opportunistic information seeking, in: C. Stephanidis, J. Jacko (Eds.), 10th International Conference on Human–Computer Interaction (HCII '03), Crete, Greece, June 2003, pp. 1131–1135.

72. P. Janecek, P. Pu, An evaluation of semantic fisheye views for opportunistic search in an annotated image collection, J. Digit. Libr. (Special issue on Information Visualisation Interfaces for Retrieval and Analysis), 4 (4) (2004) 42–56.

73. M. Hearst, Design recommendations for hierarchical faceted search interfaces, in: ACM SIGIR Workshop on Faceted Search, 2006.

74. S. Kosinov, S. Marchand-Maillet, Overview of approaches to semantic augmentation of multimedia databases for efficient access and content retrieval, in: Proceedings of the 1st International Workshop on Adaptive Multimedia Retrieval (AMR 2003), pp. 19–35, Hamburg, 2003.

75. A. Schreiber, B. Dubbeldam, J. Wielemaker, B. Wielinga, Ontology-based photo annotation, IEEE Intell. Syst. 16 (3) (2001) 66–74.

76. C. Chen, K. Börner, Top ten problems in visual interfaces of digital libraries, in: K. Börner, C. Chen (Eds.), Visual Interfaces to Digital Libraries, vol. LNCS 2539, pp. 227–232, Springer Verlag, 2002.

77. W. Croft, What do people want from information retrieval? D-Lib Mag. November 1995. Available electronically from http://www.dlib.org/dlib/november95/llcroft.htm.

78. A. Rorissa, P. Clough, T. Deselaers, Exploring the relationship between feature and perceptual visual spaces, J. Am. Soc. Inf. Sci. Technol. (JASIST), 58 (10) (2007) 1401–1418.

79. Y. Ivory, M. Hearst, The state of the art in automating usability evaluation of user interfaces, ACM Comput. Surv. 33 (4) (2001) 470–516.

80. K. Rodden, W. Basalaj, D. Sinclair, K. Wood, Does organization by similarity assist image browsing? in: ACM-CHI'01, Seattle, 2001, pp. 190–197.

81. J.A. Black, F. Gamal, P. Sethuraman, A method for evaluating the performance of content-based image retrieval systems based on subjectively determined similarity between images, in: Int. Conference on Image and Video Retrieval (CIVR'2002), vol. LNCS 2383, 2002, pp. 356–366.

82. J. Urban, J.M. Jose, Evaluating a workspace's usefulness for image retrieval, Multimedia Syst. 12 (4–5) (2006) 355–373.

83. T. Combs, B. Bederson, Does zooming improve image browsing? in: Proceedings of Digital Library (DL 99), 1999, pp. 130–137.

84. J.J. Rocchio, Relevance feedback in information retrieval, in: G. Salton (Ed.), The SMART Retrieval System, Prentice Hall, New Jersey, 1971, pp. 456–484.

85. E.Y. Chang, B. Li, G. Wu, K. Go, Statistical learning for effective visual information retrieval, in: Proceedings of the IEEE International Conference on Image Processing, pp. 609–612, 2003.

86. X. Zhou, T. Huang, Small sample learning during multimedia retrieval using biasmap, in: Proceedings of the IEEE Conference on Pattern Recognition and Computer Vision, CVPR'01, vol. 1, Hawaii, 2004, pp. 11–17.

87. X. Zhou, A. Garg, T. Huang, A discussion of nonlinear variants of biased discriminant for interactive image retrieval, in: Proceedings of the 3rd Conference on Image and Video Retrieval, CIVR'04, 2004, pp. 353–364.

88. R. Yan, A. Hauptmann, R. Jin, Negative pseudo-relevance feedback in content-based video retrieval, in: Proceedings of ACM Multimedia (MM2003), Berkeley, USA, pp. 343–346, 2003.

89. J. Kittler, M. Hatef, R. Duin, J. Matas, On combining classifiers, IEEE Trans. Pattern Anal. Mach. Intell. 20 (3) (1998) 226–239.

90. R. Duin, The combining classifier: to train or not to train? in: Proceedings of the 16th International Conference on Pattern Recognition, ICPR'02, vol. 2, IEEE Computer Society Press, Quebec City, 2004, pp. 765–770.

91. Y. Wu, E.Y. Chang, K.-C. Chang, J. Smith, Optimal multimodal fusion for multimedia data analysis, in: Proceedings of ACM International Conference on Multimedia, pp. 572–579, New York, 2004.

92. W.H. Hsu, S.-F. Chang, Generative, discriminative, and ensemble learning on multi-modal perceptual fusion toward news video story segmentation, in: ICME, Taipei, Taiwan, June 2004.

93. M. Wang, X.-S. Hua, X. Yuan, Y. Song, L.-R. Dai, Optimizing multigraph learning: towards a unified video annotation scheme, in: MULTIMEDIA '07: Proceedings of the 15th International Conference on Multimedia, ACM Press, New York, NY, USA, 2007, pp. 862–871.

94. T.-T. Pham, N.E. Maillot, J.-H. Lim, J.-P. Chevallet, Latent semantic fusion model for image retrieval and annotation, in: CIKM '07: Proceedings of the Sixteenth ACM Conference on Information and Knowledge Management, ACM Press, New York, NY, USA, 2007, pp. 439–444.

95. N. Oza, R. Polikar, J. Kittler, F. Roli, Multiple classifier systems, Lecture Notes in Computer Science, vol. 3541, Springer, 2005.

96. S.C. Hoi, M.R. Lyu, A multi-modal and multi-level ranking scheme for large-scale video retrieval, IEEE Trans. Multimedia 10 (4) (2008) 607–619.

97. J. Howe, The rise of crowdsourcing, Wired Mag. 14 (06) (2006) 1–4.

98. Netflix. The Netflix Prize. <http://www.netflixprize.com/> 2006. Accessed August 2009.

99. C.M. University, reCAPTCHA. <http://recaptcha.net/>, 2007. Accessed August 2009.

100. H. Müller, W. Müller, D.M. Squire, S. Marchand-Maillet, T. Pun, Long-term learning from user behavior in content-based image retrieval Technical report, Université de Genève, 2000.

101. L. Wenyin, S. Dumais, Y. Sun, H. Zhang, M. Czerwinski, B. Field, Semi-automatic image annotation, in: Proceedings of International Conference on HCI, pp. 326–333, 2001.

102. D. Heisterkamp, Building a latent-semantic index of an image database from patterns of relevance feedback, in: Proceedings of 16th International Conference on Pattern Recognition, pp. 134–137, 2002.

103. I. Ruthven, M. Lalmas, A survey on the use of relevance feedback for information access systems, Knowl. Eng. Rev. 18 (2) (2003) 95–145.

104. C. Anderson, The long tail, Wired Mag. 12 (10) (2004) 170–177.

105. J. Fournier, M. Cord, Long-term similarity learning in content-based image retrieval, in: Proceedings of the international conference on image processing, 2002.

106. M. Li, Z. Chen, H. Zhang, Statistical correlation analysis in image retrieval, Pattern Recognition 35 (12) (2002) 2687–2693.

107. M. Koskela, J. Laaksonen, Using long-term learning to improve efficiency of content-based image retrieval, In International Workshop on Pattern Recognition in Info. Sys., Angers, France, pp. 72–79.

108. X. He, O. King, W.-Y. Ma, M. Li, H.-J. Zhang, Learning a semantic space from user's relevance feedback for image retrieval, IEEE Trans. Circuits Sys. Video Technol. 13 (1) (2003) 39–48.

109. M. Cord, P. H. Gosselin, Image retrieval using long-term semantic learning, in: IEEE International Conference on Image Processing, pp. 2909–2912, 2006.

110. N. Craswell, M. Szummers, Random walks on the click graph, in: Proceedings of SIGIR 2007, pp. 239–246, 2007.

111. E. Pampalk, Islands of music – analysis, organization, and visualization of music archives, J. Austrian Soc. Artif. Intell. 22 (3) (2003) 20–23.

112. O. de Rooij, C.G.M. Snoek, M. Worring, Query on demand video browsing 2007, in: ACM Multimedia'07, Augsburg, Germany, pp. 811–814.

113. G. Ciocca, R. Schettini, Hierarchical browsing of video key frames, in: Proceeding of the EU Conference on Information Retrieval (ECIR'2007), 2007, pp. 691–694.

Chapter 15

Modelling Interest in Face-to-Face Conversations from Multimodal Nonverbal Behaviour

Daniel Gatica-Perez

Idiap Research Institute, Martigny, Switzerland

15.1 INTRODUCTION

Many readers can likely recall having seen young children literally jumping off their seat when they meet somebody they specially like. Many readers might also have observed the same children being mesmerised, almost still, when somebody or something catches their full attention. These are examples of interest, a fundamental internal state related to many human processes – including imagination, creativity, and learning – that is known to be revealed by nonverbal behaviour expressed through voice, gestures and facial expressions [1, 2] and that has recently been added to the research agenda on multimodal signal processing for human computing.

Dictionaries define interest as 'a state of curiosity or concern about or attention to something: an interest in sports; something, such as a quality, subject, or activity, that evokes this mental state' (The American Heritage Dictionary of the English Language) or as 'a feeling that accompanies or causes special attention to an object or class of objects; something that arouses such attention' (Merriam-Webster). In this chapter, which is focused on face-to-face conversations, the term interest is used to designate people's internal states related to the degree of engagement displayed, consciously or not, during social interaction. Such engagement can be the result of many factors, ranging from interest in the theme of a conversation, attraction to the interlocutor, and social rapport. Displays of social interest through nonverbal cues have been widely studied in social psychology and include mimicry [3, 4] (an imitation phenomenon displayed through vocal cues but also via body postures and mannerisms, and facial expressions), elevated displays of speaking and kinesic activity and higher conversational dynamics. In a conversation, interest can be expressed both as a speaker and as a listener. As a speaker, an interested person often increases both voice and body activity. The case of attraction could also involve persisting gaze. As a listener, an interested person would often show attention, expressed, e.g., via intense gaze, diminished body motion and backchannels. Mimicry would appear while playing both roles. The degree of interest that the members of a dyad or a group collectively display during their interaction could be used to extract important information. This could include inferring whether a brief interaction has been interesting to the participants and segmenting a long interaction (e.g., a group meeting at work) into periods of high and low interest. Interest level categories could therefore be used to index and browse conversations involving oneself and in some contexts involving others (e.g., at work) where segments of meetings in which participants of a team were highly engaged could be of interest to other team members who had not had the chance to attend the meeting.

This chapter briefly reviews the existing work on automatic modelling of interest in face-to-face interactions, discussing research involving both dyads and groups, and focuses on discussing the multimodal cues and machine learning models that have been used for detection and recognition of interest and related concepts. The domain is relatively new and therefore poses a considerable number of research challenges in multimodal signal processing. From a larger perspective, interest is one of many aspects that are currently studied in social computing, the computational analysis of social behaviour from sensor data [5–7].

The rest of this chapter is organised as follows. Section 15.2 summarises the various computational perspectives related to interest modelling that have been addressed in the literature. Section 15.3 reviews work on conversational interest modelling from audio nonverbal cues. Section 15.4 reviews the emerging work on conversational interest modelling from audiovisual cues. Section 15.5 discusses other research investigating problems related to interest modelling. Finally, Section 15.6 offers some concluding remarks. Parts of the material presented in this chapter have been adapted from [7].

15.2   PERSPECTIVES ON INTEREST MODELLING

Although other authors have advocated for a distinction between interest and several other related concepts like engagement or attraction [8], given the relatively small number of existing works in this domain, a presentation under a generic term was chosen to facilitate a comparative discussion. The literature on computational modelling of interest in face-to-face conversations can be categorised according to different perspectives (see also Figure 15.1):

1. *Interaction type*: Dyads, small groups and large groups have all been analysed in the literature.
2. *Processing units*: Existing works have considered the units of analysis to be (1) speech utterances by individuals, (2) interaction segments (not necessarily aligned with speech utterances) and (3) whole interactions.
3. *Target tasks*: Depending on the processing units, the target tasks have included (1) classification of presegmented speech utterances into a small set of interest-level classes (e.g., high or low interest);

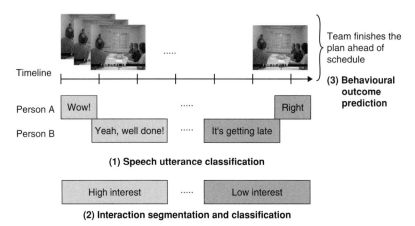

FIGURE 15.1   Interest modelling tasks for an interacting group: (1) classification of presegmented individual speech utterances as corresponding to high interest (left, light gray) or low interest (right, dark gray); (2) segmentation and classification of meeting segments as high or low interest (left or right respectively); (3) prediction of behavioural outcomes that relate to interest level (bracket on the right in the example).

(2) automatic segmentation and classification of interaction segments into interest-level classes and (3) prediction of concrete, interest-related behavioural outcomes (e.g., mutually interested people exchanging business cards after discussing at a conference), which often results in binary classification tasks. Cases 1 and 2 require manual annotation of interest-level classes, which is commonly derived from first or third-party human judgments. Case 3, on the other hand, can use the interaction outcomes themselves as annotation. In most cases, the occurrence of high interest might be an infrequent situation, which results in imbalanced data sets for learning statistical models.

4. *Single versus multimodal cues*: Speech (captured by close-talk or distant microphones) is the predominant modality in conversations and has been the most commonly investigated. A few works, however, have studied the possibility of integrating other modalities: vision from cameras or motion from wearable sensors.

The research reviewed in this chapter is summarised in Table 15.1. Examples of some of the data used in the discussed methods appear

TABLE 15.1 Research on automatic modelling of conversational interest. NVB, nonverbal behaviour; A, audio; V, video; M, body motion. ICSIMR, International Computer Science Institute Meeting Recording; MIT, Massachusetts Institute of Technology; MR, meeting recording; MMR, Multimodal Meeting Manager

Ref.	Scenario and Task	Data	NVB
[9]	4- to 8-person meetings; relation between prosodic cues and hot spots for utterances	ICSI MR corpus; 88 speech utterances from 13 meetings	A
[10]	5- to 8-person meetings; relation between hot spots and dialogue acts for utterances	ICSI MR corpus; 32 meetings; approximately 32 h	A
[11]	5-person meeting; classification of utterances as emphasised/neutral	ICSI MR corpus; 1 meeting; 22 min; 861 utterances	A
[12]	5- to 8-person meetings; classification of speech 'spurts' as agreement/ disagreement	ICSI MR corpus; 7 meetings	A
[13]	dyadic speed dates; prediction of matches of mutually interested people	MIT data; 60 five-min meetings	A
[8]	dyadic interaction; classification of short conversations as high/low interest	MIT data; 100 three-min conversations	A
[14]	9-person meetings; manual annotation of individual interest level	MIT data; 1 one-h meeting	A
[15]	4-person meetings; segmentation and classification of high/neutral group interest	M4 corpus; 50 five-min meetings;	A,V
[16]	113 and 84 conference attendees; bookmarking of dyadic encounters (high interest)	MIT data; 1 day (approximately 8 h) in each case	A, M

The investigated nonverbal behaviour includes audio, video and body motion cues

FIGURE 15.2   Scenarios and data for estimation of interest in face-to-face conversations: (a) ICSI Meeting Recording corpus [9]. (b) MIT speed dating corpus [13]. (c) M4 (MultiModal Meeting Manager) corpus [15]. (d) MIT conference corpus [16]. (e) Augmented Multiparty Interaction corpus [17]. All pictures are reproduced with permission.

in Figure 15.2. The next two sections review the existing work based on the use of single and multiple perceptual modalities, respectively.

15.3   COMPUTING INTEREST FROM AUDIO CUES

Most existing work on automatic interest modelling has focused on the relations between interest (or related concepts) and the speech modality, using both verbal and nonverbal cues. Wrede and Shriberg [9, 10] introduced the notion of hot spots in group meetings, defining it in terms of participants highly involved in a discussion and relating it to the concept of activation in emotion modelling, i.e., the 'strength of a person's disposition to take action' [18]. They used data from the International Computer Science Institute (ICSI) Meeting Recording (MR) corpus [19] containing four- to eight-person conversations, close-talk microphones and speech utterances as the basic units. In [9], defining a hot spot utterance as one in which a speaker sounded 'especially interested, surprised or enthusiastic about what is being said, or he or she could express strong disagreement, amusement, or stress' [10], the authors first developed an annotation scheme that included three categories of involvement (*amused*, *disagreeing* and *other*), one *not specially involved* category and one *I don't know* category, which human annotators used to label utterances based as much as possible on the acoustic information (rather than the content) of each utterance. This study found that human annotators could reliably perceive involvement at the utterance level (a Kappa inter-annotator analysis produced a value of 0.59 in discriminating between involved and noninvolved utterances, and lower values for the multicategory case). This work also studied a number of prosodic cues related to the energy of voiced segments and the fundamental frequency (F0) aggregated over speech utterances, computed from individual close-talk microphones. Based on a relatively small number of speech utterances, the authors found that a number of these features (mainly those derived from F0) appear to be discriminating of involved versus noninvolved utterances. No experiments for automatic hot spot classification from single or multiple features were reported.

In subsequent work [10], the same authors extended their study to analyse the relation between hot spots and dialogue acts (DAs), which indicate the function of an utterance (question, statement, backchannel, joke, acknowledgement, etc.). The study used 32 meetings where the annotation of involvement was done by one annotator continuously listening to a meeting and using the same categories as in [9] (*amused*, *disagreeing*, *other* and *non-involved*). In this larger corpus, the authors found that a rather small proportion of utterances (about 2%) corresponded to involved utterances and also found a number of trends between DA categories and involvement categories (e.g., joke DAs occur often for amused involvement, and backchannels do so for non-involvement).

In a related line of work, Kennedy and Ellis [11] addressed the problem of detecting emphasis or excitement of speech utterances in meetings from prosodic cues, acknowledging that this concept and emotional involvement might be acoustically similar. They first asked human annotators to label utterances as *emphasised* or *neutral*, as they listened to 22 min of a five-person meeting, and found that people could reliably identify emphasised utterances (full agreement across five annotators in 62% of the data and 4 of 5 agreement in 84%) but also that the number of emphasised frames is low (about 15%). They later used a very simple approach to measure emphasis based on the assumption that heightened pitch corresponds to emphasis and using pitch and its aperiodicity computed with the Yin pitch estimator as cues [20], from signals coming from individual close-talk microphones. A basic pitch model was estimated for each speaker, to take into account each person's pitch distribution, and a threshold-based rule was established to distinguish higher pitch for frames and utterances. After eliminating very short noisy speech segments, the method produced a performance of 24% precision, 73% recall and 92% accuracy for utterances with high agreement in human judgement of emphasis.

Other existing works can also be related to the detection of high-interest segments of conversations. As one example, Yu et al. [21] also attempted to detect conversational engagement but used telephone rather than face-to-face, dyadic conversations for experiments. As another example, Hillard et al. [12] proposed a method to recognise a specific kind of interaction in meetings (agreement vs. disagreement) that is likely related to high interest. Using seven meetings from the ISCI corpus, the work used speech 'spurts' (speech intervals with no pauses greater than 0.5 s) as processing units, which are to be classified as *positive* or *backchannel* (corresponding to the agreement class), *negative* (the disagreement class) and *other*. On a subset of the data, about 15% of the spurts corresponded to either agreement or disagreement. For classification, both prosodic cues (including pause duration, fundamental frequency and vowel duration) and word-based features (including the total number of words and the number of 'positive' and 'negative' keywords) were used in a learning approach that made use of unlabelled data. In the three-way classification task, the authors found that clean speech transcripts performed the best (which is not surprising given that the manual annotation of spurts took their content into account), and that prosody produced promising performance (with classification accuracy similar to the option of using keywords and noisy automatic speech recognition (ASR) transcripts), although fusing ASR transcripts and prosody did not improve performance.

The work by Pentland and collaborators has also dealt with the estimation of interest and related quantities [6, 13, 14, 22, 23], in both dyadic and group cases. One key feature of this line of work is that it has often studied social situations with concrete behavioural outcomes (e.g., people declaring common attraction in a speed dating situation or people exchanging business cards at a conference as a sign of mutual interest), which substantially reduces the need for third-party annotation of interest. Madan et al. studied a speed-dating dyadic scenario for prediction of attraction (that is, romantic or friendly interest) between different-gender strangers in 5-min encounters [8, 13]. In this scenario, participants interact with several randomly assigned 'dates' and introduce each other for a short period of time and privately decide whether they are interested in seeing this person again (labelling their interaction partner as a 'yes' or 'no' for three cases: *romantically attracted*, *interested in friendship* or *interested in business*). Matches are then found by a third person at the end of the session, when two interaction partners agree on their mutual interest in exchanging contact information. The authors recorded 60 five-min speed dates with audio-only sensors (directional microphones). Four nonverbal audio cues, dubbed activity level, engagement, stress and mirroring were extracted [5]. The activity level is the z-scored percentage of speaking time computed over speaking voiced segments. Engagement is the z-scored influence a person has on the turn-taking patterns of the others (influence itself is computed with a hidden Markov model (HMM)). Stress is the z-scored sums of the standard deviations of the mean-scaled energy, fundamental frequency and spectral entropy of voiced segments. Finally, mirroring is the z-scored frequency of short utterance (less than 1-s long) exchanges. For the *attracted* category, the authors observed that women's nonverbal cues were correlated to both female and male attraction (*yes*) responses (activity level being the most predictive cue), whereas men's nonverbal cues had no significant correlation with attraction responses. Other results also showed some other cues to be correlated with the *friendship* or *business* responses. An additional analysis of the results, along with pointers for implementation of the used nonverbal cues, can be found in [23]. Madan et al. also used these cues in different combinations as input to standard classifiers like linear classifiers or support vector machines (SVM) and obtained promising performance (70–80% classification accuracy).

In another dyadic case, Madan and Pentland targeted the prediction of interest-level (*high* vs. *low*) in 3-min conversations between same-gender people discussion about random topics [8, 22]. Twenty participants of both genders were first paired with same-gender partners. Each pair participated in 10 consecutive 3-min conversations and ranked their interest on a 10-point scale after each encounter. In [8], using the same set of features as for the speed dating case and a linear SVM classifier, the best features could correctly classify binary interest levels with 74% accuracy for males, whereas different behaviour was observed for females, and no results were reported for automatic classification.

Pentland et al. have also investigated multiparty scenarios. In early work, Eagle and Pentland investigated the group conversation case, where the interest level in the ongoing conversation was manually introduced by users in a mobile device [14], from which a group interest level could be inferred via averaging. Although the device was designed so that the annotation process would not be over distracting, there is still a cognitive load cost associated to this interactive task.

15.4 COMPUTING INTEREST FROM MULTIMODAL CUES

Even though it is known that interest in conversations is displayed through vocal and kinesic nonverbal behaviour, few works up to date have studied the use of multiple modalities for interest estimation, using joint data captured by microphones, cameras or other sensors.

In the context of small group meetings, Gatica-Perez et al. presented in [15] an investigation of the performance of audio-visual cues on discriminating high versus neutral group interest-level segments, i.e., on estimating single labels for meeting segments, much like hot spots, using a supervised learning approach that simultaneously produces a temporal segmentation of the meeting and the binary classification of the segments into high or neutral interest-level classes. Experiments were conducted on a subset of the MultiModal Meeting Manager (M4) data corpus [24], consisting of 50 five-min four-person conversations recorded with three cameras and 12 microphones (including four lapels and one eight-microphone array). These meetings were recorded based on turn-taking scripts, but otherwise the participants behaviour was reasonably natural with respect to emotional engagement. Regarding human annotation of interest, unlike other works discussed in this chapter [9–11], which used speech utterances to produce the ground-truth, the work in [15] used interval coding [25] and relied on multiple annotators that continuously watched the meeting and labelled 15-s intervals in a 5-point scale. The ground truth (meeting segments labelled either *neutral interest* or *high interest*) was produced after an analysis of inter-annotator agreement which showed reasonable agreement and later used for training and evaluation purposes (about 80% of the frames were labelled as neutral). The investigated nonverbal cues included audio cues derived from lapel microphones (pitch, energy, speaking rate) and from microphone arrays (speech activity estimated by the steered power response phase transform. Visual nonverbal cues were also extracted for each participant by computing skin-colour blobs motion and location, as a rough proxy for head and body motion and pose. Two HMM recognition strategies were investigated [26]: early integration, where all cues were synchronised and concatenated to form the observation vectors; and multistream HMMs, in which the audio and the visual modalities are used separately to train a single-model HMM, and then both models are fused at the state level to do inference (decoding). Various combinations of audio, visual, and multimodal cues and HMM models were investigated. The performance of automatic segmentation and segment labelling was evaluated at the frame-level based on a convex combination of precision and recall (instead of using a more standard measure similar to the Word Error Rate in speech recognition that might not be meaningful when recognising binary sequences). Overall, the results were promising (some of the best reported precision/recall combinations were 63/85 and 77/60) and indicated that combining multiple audio cues outperformed the use of individual cues, that audio-only cues outperformed visual-only cues and that audio-visual fusion brought benefits in some precision/recall conditions, outperforming audio-only cues, but not in others.

In a different scenario, Gips and Pentland investigated the conference case, where large groups of attendees participate and multiple brief conversational exchanges occur [16, 22]. A sensor badge worn by the attendees recorded audio, motion from accelerometers and proximity to other badges via infrared. Additionally, people could bookmark other attendees they had interacted with by pressing a button, in the understanding that the contact details of bookmarked people would be automatically made available after the conference. In this case, the task was to predict for what encounters people bookmark their conversation partner. Two data sets were collected: one involving 113 people in a sponsor conference and another involving 84 participants recorded 6 months later. Using a set of 15 basic features derived from the accelerometer and microphone (mean and standard deviation of the amplitude and difference of the raw signals computed over time windows), the authors found that both audio and motion cues were significantly correlated with bookmarks (specially with the standard deviation cues). Using a quadratic classifier and the subset of the six most correlated cues resulted in 82.9 and 74.6% encounter classification accuracy (*bookmarked* vs. *nonbookmarked*) for each of the two data sets.

15.5   OTHER CONCEPTS RELATED TO INTEREST

As discussed in the introduction, there is a clear relation between conversational interest and attention [6]. The automatic estimation of attention could thus be important as a cue for interest modelling. It is known that listeners manifest attention by orienting their gaze to speakers, who in turn gaze to indicate whom they address and are interested in interacting with [27]. As pointed out by Knapp and Hall, people 'gaze more at people and things perceived as rewarding' and 'at those whom they are interpersonally involved' [1] (pp. 349 and 351), and this, in the context of conversations, includes people of interest. As two examples of the above, increased gaze often occurs in cases of physical attraction [1], and mutual liking has been reported to be related to gaze in dyadic discussions [28].

Estimating eye gaze in arbitrary conversational situations is a difficult problem given the difficulty in using eye trackers due to sensor setting and image resolution. Although some solutions using wearable cameras have started to appear [29], the problem of estimating gaze in conversations has been more often tackled by using head pose as a gaze surrogate. This has generated an increasing body of work [30–33] that is not reviewed here for space reasons. However, one of the most interesting aspects of current research for attention modelling is the integration of audio-visual information to estimate visual attention in conversations. In the context of group conversations, the works by Otsuka et al. [34–36] and Ba and Odobez [37] stand out as examples of models of the interplay between speaking activity and visual attention. Otsuka et al. proposed a dynamic Bayesian network (DBN) approach which jointly infers the gaze pattern for multiple participants and the conversational gaze regime responsible for specific speaking activity and gaze patterns (e.g., all participants converging onto one person or two people looking at each other) [34]. Gaze was approximated by head pose, observed either through magnetic head trackers attached to each participant [34] or automatically estimated from video [35]. Otsuka et al. later extended their model in an attempt to respond to 'who responds to whom, when and how' questions in a joint manner [36]. With somewhat similar hypotheses, Ba and Odobez proposed a DBN model for the estimation of the joint attention of group participants by using people's speaking activity as a contextual cue, defining a prior distribution on the potential visual attention targets of each participant [37]. This observation resulted in improved visual attention recognition from head orientation automatically estimated from a single camera on a subset of the Augmented Multiparty Interaction (AMI) meeting corpus, a publicly available meeting collection with audio, video, slides, whiteboard and handwritten note recordings [17] (also see Figure 15.2.)

Listening is a second conversational construct clearly related to attention. Listening is in principle a multimodal phenomenon, and some works have started to investigate computational models. Heylen et al. [38] presented research towards building a Sensitive Artificial Listener, based on the manual annotation of basic nonverbal behaviours displayed by listeners in group conversations, including gaze, head movements and facial expressions extracted from the AMI corpus.

Finally, there is recent body of work by Pentland et al. that is beginning to investigate the recognition of longer term phenomena in real-life organisational settings involving large groups. These organisational behaviour phenomena, although clearly distinct from the concept of interest as discussed here, are nevertheless related to the aggregation of states of human interest over time. More specifically, this research has examined the correlation between automatically extracted nonverbal cues and concepts like workload, job satisfaction, and productivity in banks [39, 40] and hospitals [41] and team performance and individual networking performance in professional gatherings [42]. In all cases, sensing is done through sociometers, i.e., wearable devices capable of measuring a number of nonverbal cues including physical proximity, actual face-to-face interaction, body motion and audio. Overall, this is an example of the complex sociotechnical research that will continue to appear in the future regarding social behaviour analysis and that might make use of interest or similar concepts as mid-level representations for higher social inference.

15.6   CONCLUDING REMARKS

This chapter has presented a concise review of work related to interest modelling in face-to-face conversations from multimodal nonverbal behaviour. The discussion in the previous sections highlights the facts that this domain is still emerging and that many opportunities lie ahead regarding the study of other automatic nonverbal cues that are correlated with displays of interest (importantly, from the visual modality), the design of new multimodal integration strategies and the application of cues and models to other social scenarios. The improvement of the technological means to record real interaction, both in multisensor spaces and with wearable devices, is opening the possibility to analyse multiple social situations where interest emerges and correlates with concrete social outcomes and also to develop new applications related to self-assessment and group awareness. Given the increasing attention in signal processing and machine learning with respect to social interaction analysis, there is much to look forward to in the future regarding advances on computational modelling of social interest and related concepts.

Acknowledgments

The author thanks the support of the Swiss National Center of Competence in Research on Interactive Multimodal Information Management (IM2) and EC project Augmented Multi-Party Interaction with Distant Access (AMIDA). He also thanks Nelson Morgan (ICSI) and Sandy Pentland (MIT) for giving permission to reproduce some of the pictures presented in this chapter (in Figure 15.2).

REFERENCES

1. M.L. Knapp, J.A. Hall, Nonverbal Communication in Human Interaction, Sixth ed., Wadsworth Publishing, 2005.
2. Y.S. Choi, H.M. Gray, N. Ambady, The glimpsed world: unintended communication and unintended perception, in: R.H. Hassin, J.S. Uleman, J.A. Bargh, (Eds.), The New Unconscious, Oxford University Press, 2005.
3. T.L. Chartrand, J.A. Bargh, The chameleon effect: the perception-behavior link and social interaction, J. Pers. Soc. Psychol. 76 (6) (1999) 893–910.
4. T.L. Chartrand, W. Maddux, J. Lakin, Beyond the perception-behavior link: the ubiquitous utility and motivational moderators of nonconscious mimicry, in: R. Hassin, J. Uleman, J.A. Bargh, (Eds.), The New Unconscious. Oxford University Press, 2005.
5. A. Pentland, Socially aware computation and communication, IEEE Comput. 38 (2005) 63–70.
6. A. Pentland, Honest Signals: How they Shape Our World. MIT Press, 2008.
7. D. Gatica-Perez, Automatic nonverbal analysis of social interaction in small groups: a review, Image and Vision Computing 2009.

8. A. Madan, Thin slices of interest, Master's thesis, Massachusetts Institute of Technology, 2005.

9. B. Wrede, E. Shriberg, Spotting hotspots in meetings: human judgments and prosodic cues, Proceedings of the Eurospeech Conference, 2003. pp. 2805–2808.

10. B. Wrede, E. Shriberg, The relationship between dialogue acts and hot spots in meetings, Proceedings of the IEEE Automatic Speech Recognition and Understanding Workshop (ASRU), 2003, pp. 180–184.

11. L. Kennedy, D. Ellis, Pitch-based emphasis detection for characterization of meeting recordings, Proceedings of the ASRU, workshop, pp. 243–248, 2003.

12. D. Hillard, M. Ostendorf, E. Shriberg, Detection of agreement vs. disagreement in meetings: training with unlabeled data, Proceedings of HLT-NAACL Conference, 2003, pp. 34–36.

13. A. Madan, R. Caneel, A. Pentland, Voices of attraction, Proceedings of the International Conference on Augmented Cognition (AC-HCI), 2005.

14. N. Eagle, A. Pentland, Social network computing, Proceedings of the International Conference on Ubiquituous Computing (UBICOMP), 2003, pp. 289–296.

15. D. Gatica-Perez, I. McCowan, D. Zhang, S. Bengio, Detecting group interest-level in meetings, Proceedings of the IEEE International Conference on Acoustics, Speech and Signal Processing (ICASSP), 2005, pp. 489–492.

16. J. Gips, A. Pentland. Mapping human networks, Proceedings of the IEEE International Conference on Pervasive Computing and Communications, 2006, pp. 159–168.

17. J. Carletta, S. Ashby, S. Bourban, M. Flynn, M. Guillemot, T. Hain, et al., The AMI meeting corpus: a pre-announcement, Proceedings of the Workshop on Machine Learning for Multimodal Interaction (MLMI), 2005, pp. 28–39.

18. R. Cowie, E. Douglas-Cowie, N. Tsapatsoulis, G. Votsis, S. Kollias, W. Fellenz, et al., Emotion recognition in human-computer interaction, IEEE Signal Process. Mag. (2001) 18(1).

19. A. Janin, D. Baron, J. Edwards, D. Ellis, D. Gelbart, N. Morgan, et al., The ICSI meeting corpus, Proceedings of the IEEE International Conference on Acoustics, Speech and Signal Processing (ICASSP), 2003, pp. 364–367.

20. A. de Cheveigne, H. Kawahara, Yin, a fundamental frequency estimator for speech and music, J. Acoustic Soc. Am., 2001, III(4), pp. 1917–1930.

21. C. Yu, P. Aoki, A. Woodruff, Detecting user engagement in everyday conversations, Proceedings of the ICSLP, 2004, pp. 1329–1332.

22. A. Pentland, A. Madan. Perception of social interest, Proceedings of the IEEE International Conference on Computer Vision, Workshop on Modeling People and Human Interaction (ICCV-PHI), 2005.

23. W.T. Stoltzman, Toward a social signaling framework: activity and emphasis in speech, Master's thesis, Massachusetts Institute of Technology, 2006.

24. I. McCowan, D. Gatica-Perez, S. Bengio, G. Lathoud, M. Barnard, D. Zhang, Automatic analysis of multimodal group actions in meetings, IEEE Trans. Pattern Anal. Mach. Intell. 27 (3) (2005) 305–317.

25. R. Bakeman, J.M. Gottman, Observing Interaction: An Introduction to Sequential Analysis, Cambridge University Press, 1997.

26. L.R. Rabiner, B.-H. Juang, Fundamentals of Speech Recognition, Prentice-Hall, 1993.

27. C. Goodwin, Conversational Organization: Interaction Between Speakers and Hearers, vol. 11, Academic Press, New York, NY, 1981.

28. F.J. Bernieri, J.S. Gills, J.M. Davis, J.E. Grahe, Dyad rapport and the accuracy of its judgment across situations: a lens model analysis, J. Pers. Soc. Psychol. 71 (1996) 110–129.

29. B. Noris, K. Benmachiche, A. Billard, Calibration-free eye gaze direction detection with Gaussian processes, Proceedings of the International Conference on Computer Vision Theory and Applications, 2008.

30. R. Stiefelhagen, Tracking focus of attention in meetings, International Conference on Multimodal Interfaces (ICMI), 2002.

31. R. Stiefelhagen, J. Yang, A. Waibel, Modeling focus of attention for meeting indexing based on multiple cues, IEEE Trans. Neural Netw. 13 (4) (2002) 928–938.

32. S.O. Ba, J.-M. Odobez, A probabilistic framework for joint head tracking and pose estimation, Proceedings of the International Conference on Pattern Recognition (ICPR), 2004, pp. 264–267.

33. S.O. Ba, J.-M. Odobez, A study on visual focus of attention modeling using head pose, Proceedings of the Workshop on Machine Learning for Multimodal Interaction (MLMI), 2006, pp. 75–87.

34. K. Otsuka, Y. Takemae, J. Yamato, H. Murase, Probabilistic inference of multiparty-conversation structure based on markov-switching models of gaze patterns, head directions, and utterances, Proceedings of the International Conference on Multimodal Interfaces (ICMI), 2005, pp. 191–198.

35. K. Otsuka, J. Yamato, Y. Takemae, H. Murase, Conversation scene analysis with dynamic bayesian network based on visual head tracking, Proceedings of the IEEE International Conference on Multimedia (ICME), 2006, pp. 944–952.

36. K. Otsuka, J. Yamato, H. Sawada, Automatic inference of cross-modal nonverbal interactions in multiparty conversations, Proceedings of the International Conference on Multimodal Interfaces (ICMI), 2007, pp. 255–262.

37. S.O. Ba, J.M. Odobez, Multi-party focus of attention recognition in meetings from head pose and multimodal contextual cues, Proceedings of IEEE International Conference on Acoustics, Speech, and Signal Processing (ICASSP), 2008, 2221–2224.

38. D. Heylen, A. Nijholt, M. Poel, Generating nonverbal signals for a sensitive arti-
    ficial listener, Proceedings of COST 2102 Workshop on Verbal and Nonverbal
    Communication Behaviours, 2007, pp. 267–274.
39. D. Olguin Olguin, B. Weber, T. Kim, A. Mohan, K. Ara and A. Pentland, Sensi-
    ble Organizations: Technology and Methodology for Automatically Measuring
    Organizational Behavior, IEEE Trans. on Systems, Man and Cybernetics –
    Part B, Cybernetics Vol.39, No.1, Feb. 2009.
40. P.A. Gloor, D. Oster, J. Putzke, K. Fischback, D. Schoder, K. Ara, et al., Studying
    microscopic peer-to-peer communication patterns, Proc. Americas, Conference
    on Information Systems (2007).
41. D. Olguin Olguin, P.A. Gloor, A. Pentland, Capturing individual and group
    behavior with wearable sensors, Proceedings of the AAAI Spring Symposium
    on Human Behavior Modeling, 2009, pp. 68–74.
42. D. Olguin Olguin, A. Pentland, Social sensors for automatic data collection,
    Proceedings of the Americas Conference on Information Systems, 2008.

Index

327

Index

L

Language model, 17, 28
Late-integration, 37
Levenshtein distance, 132
Likelihood, 20
Listening, 322
Loss function, 10

M

Markov decision process, 79–80
Markov process, 18
Maximum a posteriori, 17
Meeting Recording corpus, 314
Monte Carlo techniques, 110
Multimodal, 1
MultiModal Meeting Manager, 314

N

Natural and spoken language
    understanding, 64
Natural language, 64
Natural language and dialogue, 63
Natural language generation, 71
Natural language processing, 63
Natural language understanding, 64
Naturalness, 44
Nonverbal behaviour, 309

O

Observations, 31
Ontology, 69

P

Paradise, 82
PDA, 122
Pen-based user input, 122
Precision and recall rates, 121

Pronunciation model, 28
Prosody, 48, 317

R

Regression, 8, 9
Rubine's algorithm, 120, 129, 130

S

Semantic parsing, 64, 68
Shape recognition, 129, 131
Simulation, 80, 84
Sketching, 134, 138, 139
SketchiXML, 120, 129, 131, 133
Source-filter, 29
Speech Act, 69
Speed dating corpus, 314
State-transition network, 66
Statistical features, 130
Support vector machines, 8, 13
Support vectors, 16
Syntactic parsing, 64
Systems, precision and recall rates, 137

T

Tablet PCs, 122
Taxonomy, 69
Temporal features, 130
Text-to-speech, 26
Transition probabilities, 19

U

Unimodal, 2
UsiXML, 129, 133

V

Visual attention, 321
Viterbi, 21, 34